Keith Wicks

illustrated by Pavel Kostal

Lerner Publications Company • Minneapolis

This edition first published 1988 by Lerner Publications Company

Library of Congress Cataloging-in-Publication Data

Wicks, Keith.
Science can be fun / by Keith Wicks; illustrated by Pavel Kostal.
p. cm. —(Do it yourself books)
Reprint. Originally published: London: Macmillan, 1982.
Summary: Provides instructions for various science projects, challenging students to discover how the world around them really works.
ISBN 0-8225-0896-6 (lib. bdg.)
1. Science—Experiments—Juvenile literature. [1. Science—Experiments. 2. Experiments.] I. Kostal, Pavel, ill. II. Title. III. Series.
Q164.W53 1988
507'.8—dc19 87-27319
CIP
AC

Manufactured in the United States of America
2 3 4 5 6 7 8 9 10 97 96 95 94 93 92 91 90 89

TOUCH

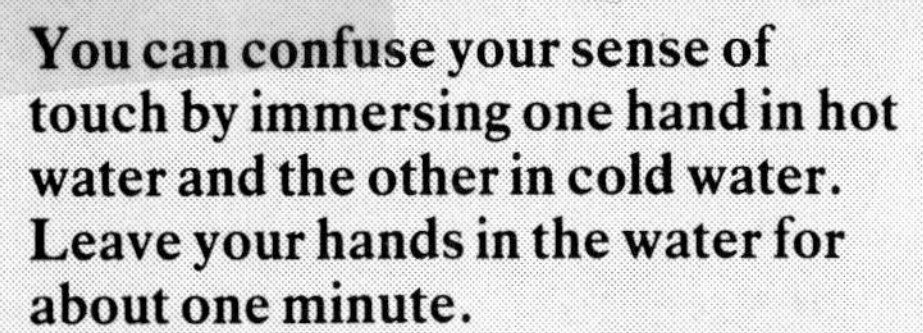

You can confuse your sense of touch by immersing one hand in hot water and the other in cold water. Leave your hands in the water for about one minute.

Now immerse both hands in warm water. This will feel cold to the hand that was in the hot water, and hot to the hand that was in the cold water.

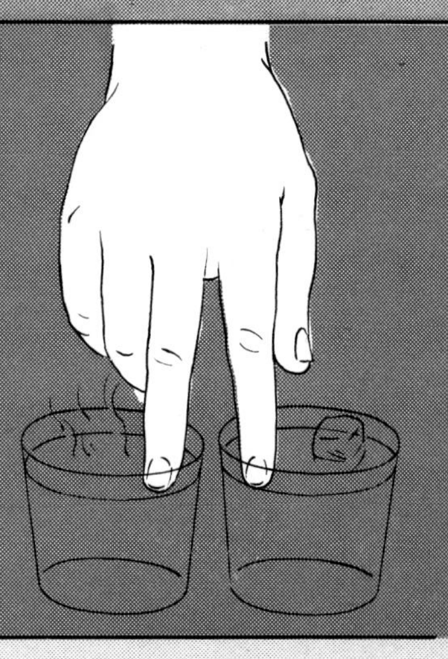

Try the test again, but this time use two fingers of one hand. Now, when you plunge both fingers into the warm water, can you detect any difference in temperature?

Place a number of objects on the ground to mark out a winding path. Allow a friend to make a mental note of where the objects are, then blindfold him. He must now try to follow the path without stepping on any obstacle. Mistakes usually occur after about eight seconds, so those who hurry may complete the test without fault.

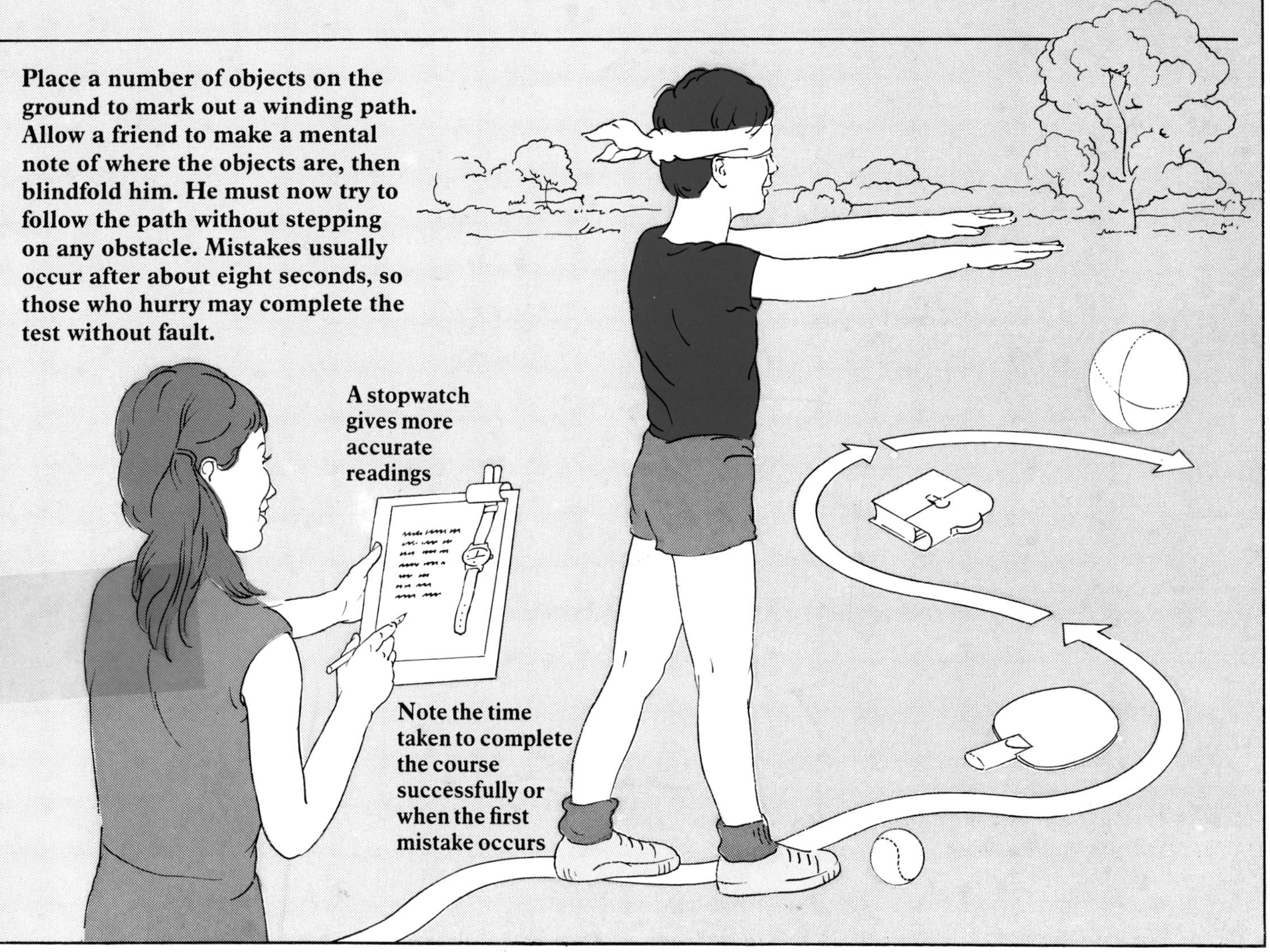

A stopwatch gives more accurate readings

Note the time taken to complete the course successfully or when the first mistake occurs

KEEPING AN AQUARIUM

Keeping a home aquarium is a simple and rewarding hobby, but care must be taken if the fish are to survive. Like us, they need a constant supply of oxygen. In natural conditions, the water surface absorbs oxygen from the air. Fish take in water through the mouth and pass it through breathing organs called gills. These absorb some of the dissolved oxygen and allow it to enter the fishes' blood-stream. The water then leaves through gills in the sides of the body.

Fish kept in a round bowl with a small opening rarely live long, as the water surface cannot absorb sufficient oxygen. This is why a fish tank with a large surface area is essential. But take care not to crowd it with too many fish.

AQUARIUM FISH

A selection of fish suitable for the cold-water aquarium are shown below and right. Goldfish are a popular choice; there are many varieties including the veiltail and fantail.

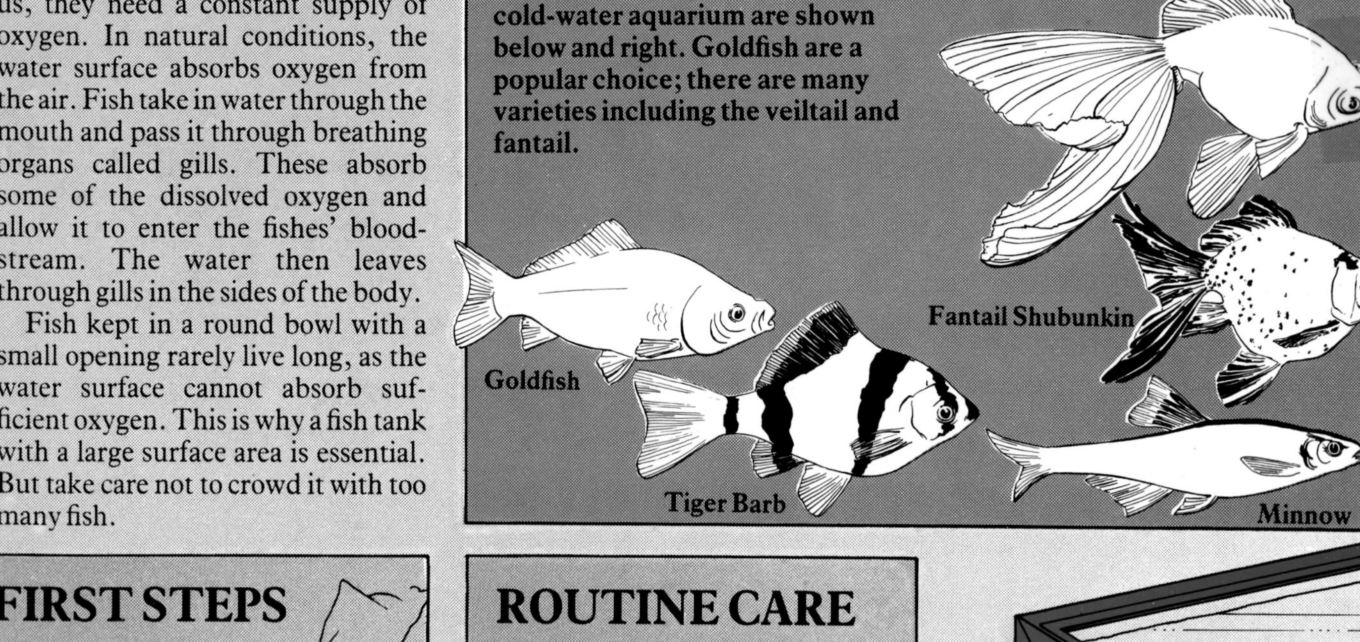

FIRST STEPS

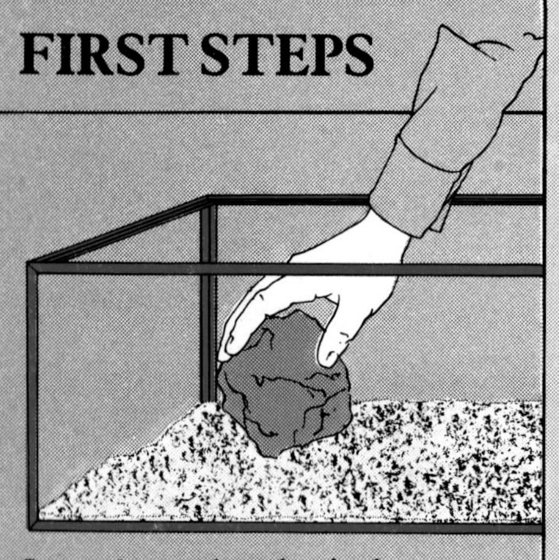

Spread gravel so that it slopes down slightly towards the front. This will simplify waste removal.

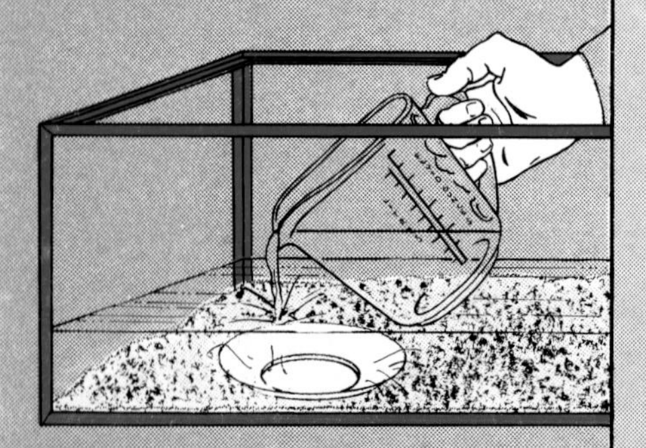

Fill the aquarium by pouring water onto a saucer. This avoids disturbing the sand and gravel and uprooting any of the water plants. Place these mainly to the back of the aquarium.

ROUTINE CARE

From time to time, a residue of uneaten food and solid waste from the fish will collect on the bottom of the aquarium. This waste material will drift down the sloping gravel and collect at the front of the tank. It should be removed promptly using a glass or plastic tube (*see below*).

Place your thumb over one end of a wide tube and position the other end over the waste material. On removing your thumb, the waste will enter the tube. Then replace your thumb and remove the tube.

It is important to keep the water in your aquarium as clear as possible. A filter system can help to do this, because it removes unwanted matter. There are many kinds of filters, some of which are fixed into the gravel at the bottom of the fish tank.

Great Ramshorn snails are of enormous benefit to the aquarium. They will eat unwanted algae that often grow on the sides of the tank. They also eat waste food material.

A cover over the tank will stop inquisitive cats from disturbing the fish. Electric lighting can be installed in the cover.

Filter

Great Ramshorn snails

Eel Grass

Brazilian Milfoil

The best place for the aquarium is under fairly strong artificial light. Minute unwanted plants called algae may infest the tank if there is too much daylight.

AQUARIUM PLANTS

Besides looking attractive, water plants serve two useful purposes. They use up some of the waste products from the fish, and they help to maintain the water's supply of dissolved oxygen.

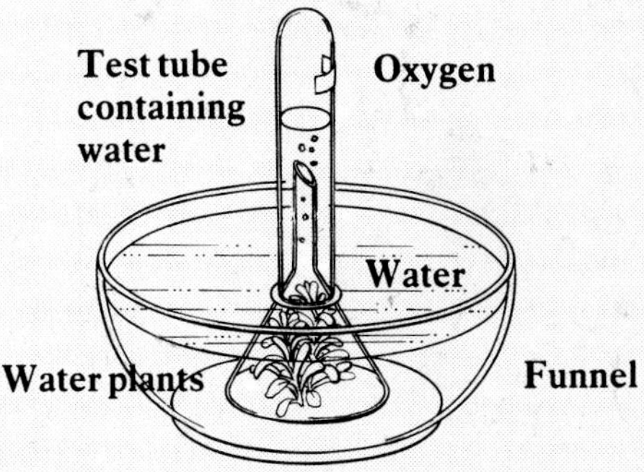

Aquarium plants give off bubbles of oxygen gas. Some of this dissolves in the water. The rest rises to the surface of the water.

You can collect some oxygen, as shown above. This gas will ignite a glowing splint.

MAGNETISM

A piece of metal that attracts other pieces of metal to it is said to be magnetized. Some metals, such as iron, tend to lose their magnetic properties quickly. Other metals, such as steel, retain these properties if magnetized. These metals are used to make permanent magnets.

All magnets have two magnetic poles – regions in which their magnetism seems to be concentrated. These are referred to as the north and south poles. Both poles will attract pieces of unmagnetized iron or steel, and the north pole of one magnet will attract the south pole of another. But two north poles or two south poles will repel each other. In a magnetic compass, the Earth's magnetism attracts a small, magnetized needle.

MAGNETIC COMPASS

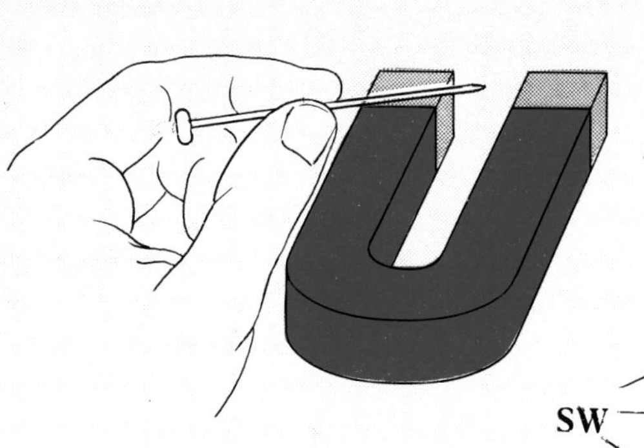

This simple compass consists of a floating cork with a magnetized steel nail through the center. Magnetize the nail by stroking it, in one direction only, with one pole of a permanent magnet.

Drive the nail through a cork an place in a shallow dish of water. Mark the points of a compass o cardboard and line up its north mark with the end of the nail that pointing north. The cardboard no shows all directions correctly.

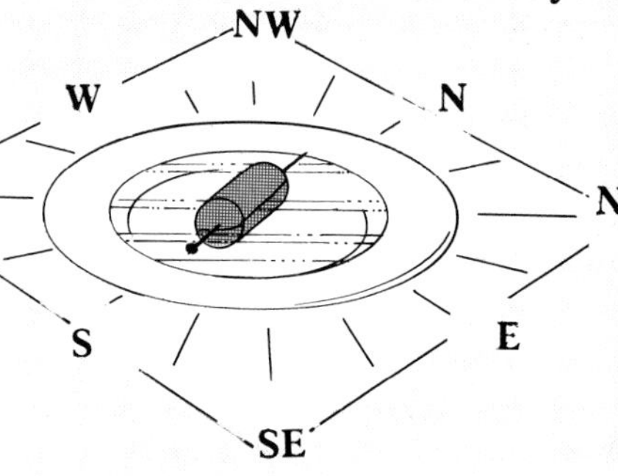

MAGNETIC FIELDS

A magnetic field is a region in which magnetism can be detected. The field around a magnet can be represented by a series of lines joining the magnet's poles.

To show the field around a bar magnet, first place a sheet of cardboard over it. Then sprinkle some iron filings onto the cardboard and tap it gently.

Right: A large steel nail can be made into an electromagnet. Cover the nail with tape, then wind on several layers of fine copper wire. This must be the type covered with insulating enamel paint or cotton.
Remove the insulation from the ends of the coil and connect, as shown, to a battery and switch.

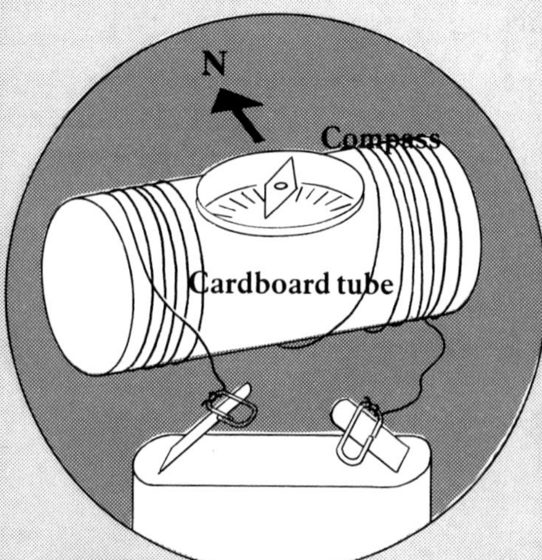

GALVANOMETER

This simple galvanometer will detect electric current. When the two ends of the coil of wire are connected to the battery, the compass needle will move.

ELECTROMAGNETS

Magnetism can be detected around a wire carrying an electric current. This effect is called electromagnetism. If the current is passed through a coil of wire, the magnetism increases in strength. Winding the coil on a piece of iron or steel increases the effect still further. Such an arrangement is called an electromagnet.

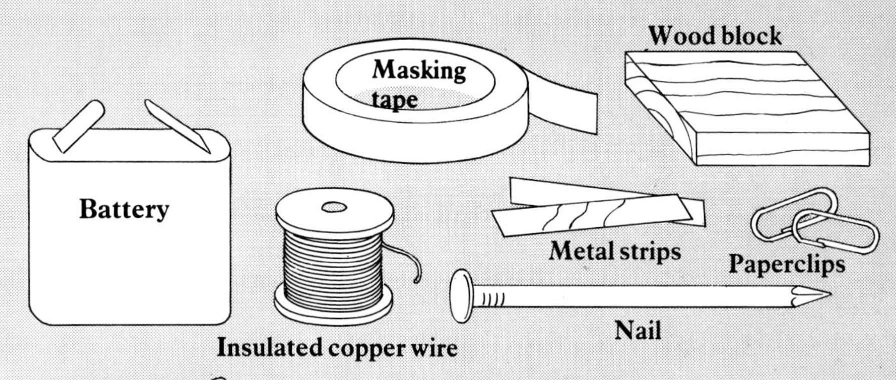

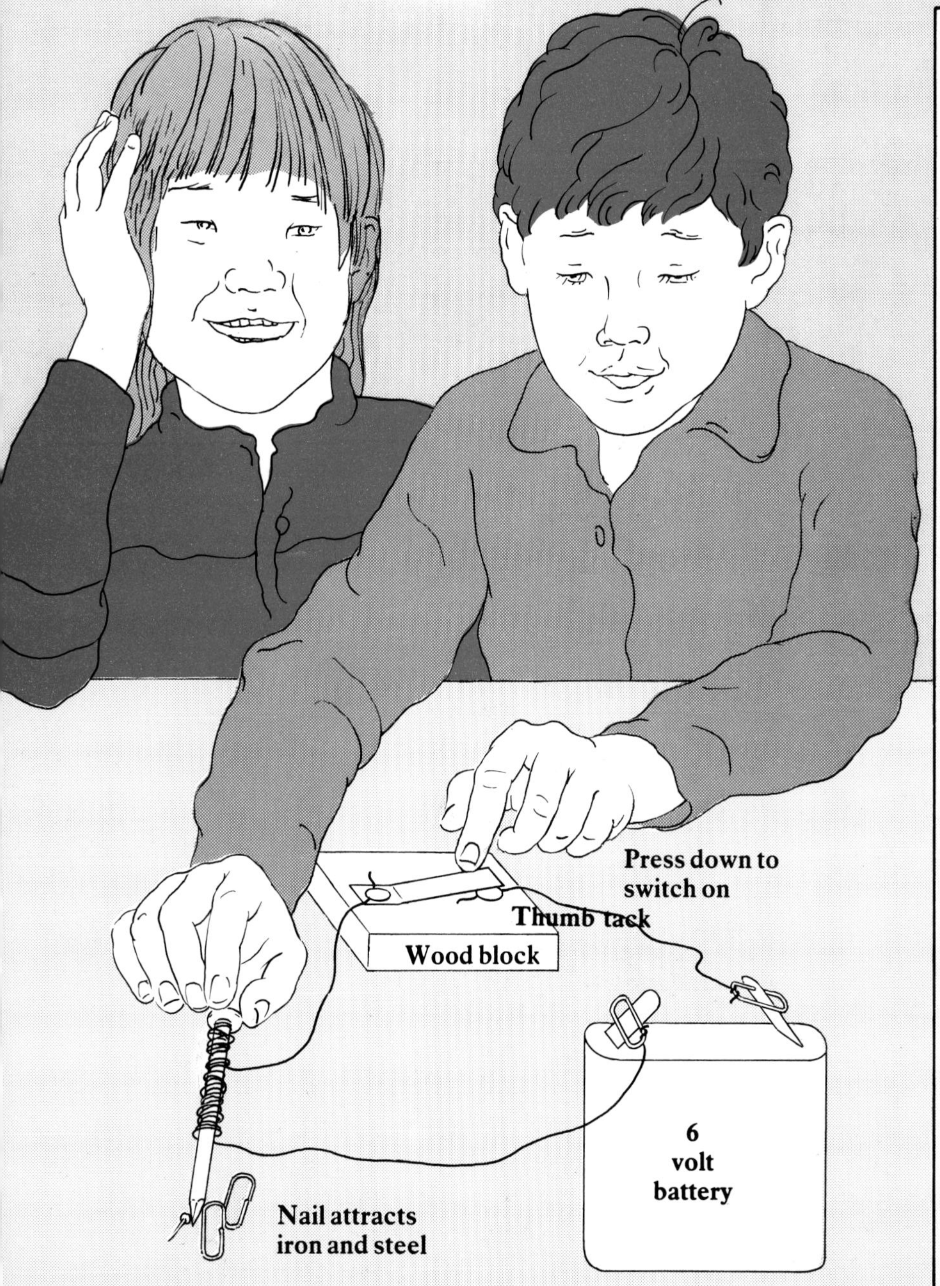

BUZZER

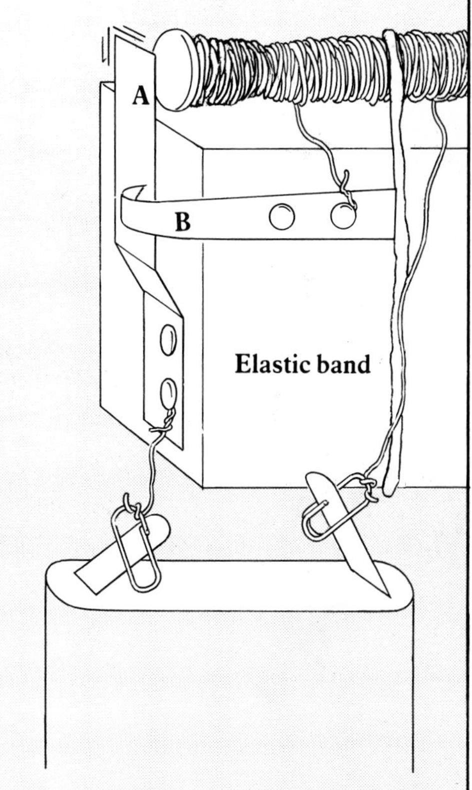

Use the electromagnet to make this buzzer. Current flows through the coil via the metal strips. This turns the nail into a magnet which attracts strip A. As strip A moves, it breaks contact with strip B. This cuts the current off, and strip A springs back again. The process repeats itself rapidly, causing a loud buzzing noise.

RADIO RECEIVERS

In the early days of radio, people listened to programs on crystal sets. A typical crystal set has no battery. It relies entirely on the energy of the radio signal it receives. So, in order to obtain a strong signal, a long aerial wire is used. Radio waves make an electric current flow down the wire through the set to a metal pipe or rod stuck in the ground.

Since the aerial picks up a mixture of radio signals, the set must have a tuning circuit to select the one required. A device called a crystal diode converts the selected signal into a form that can be reproduced on earphones. Compared with a transistor set, the crystal receiver gives a weak signal. However, it costs absolutely nothing to operate.

WHAT YOU NEED

These items, many from an electrical store, are all you need to make the crystal sets shown opposite. You can also make a radio from a kit found at an electrical store. For tuning, one set uses the metal strip. The other set is tuned with the capacitor.

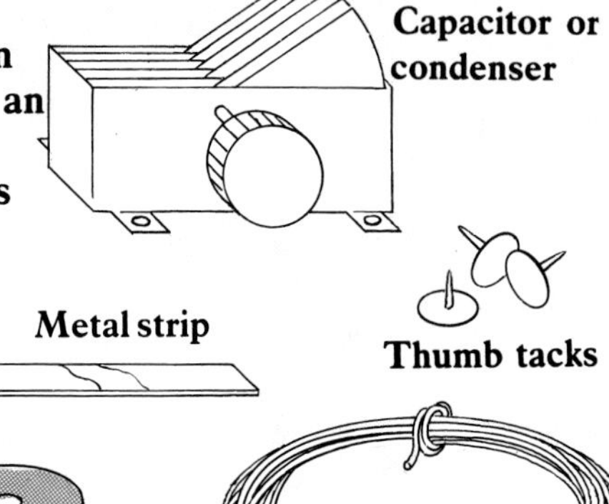

Headphones

Crystal diode or other high-speed diode

Insulated copper wire

Plastic-covered wire (for aerial and earth connection

Connect the headphones, aerial and earth wire to the set. Then adjust the tuning control until you hear a broadcast. Adding more turns to the tuning coil will enable you to receive stations broadcasting on longer wavelengths (lower frequencies).

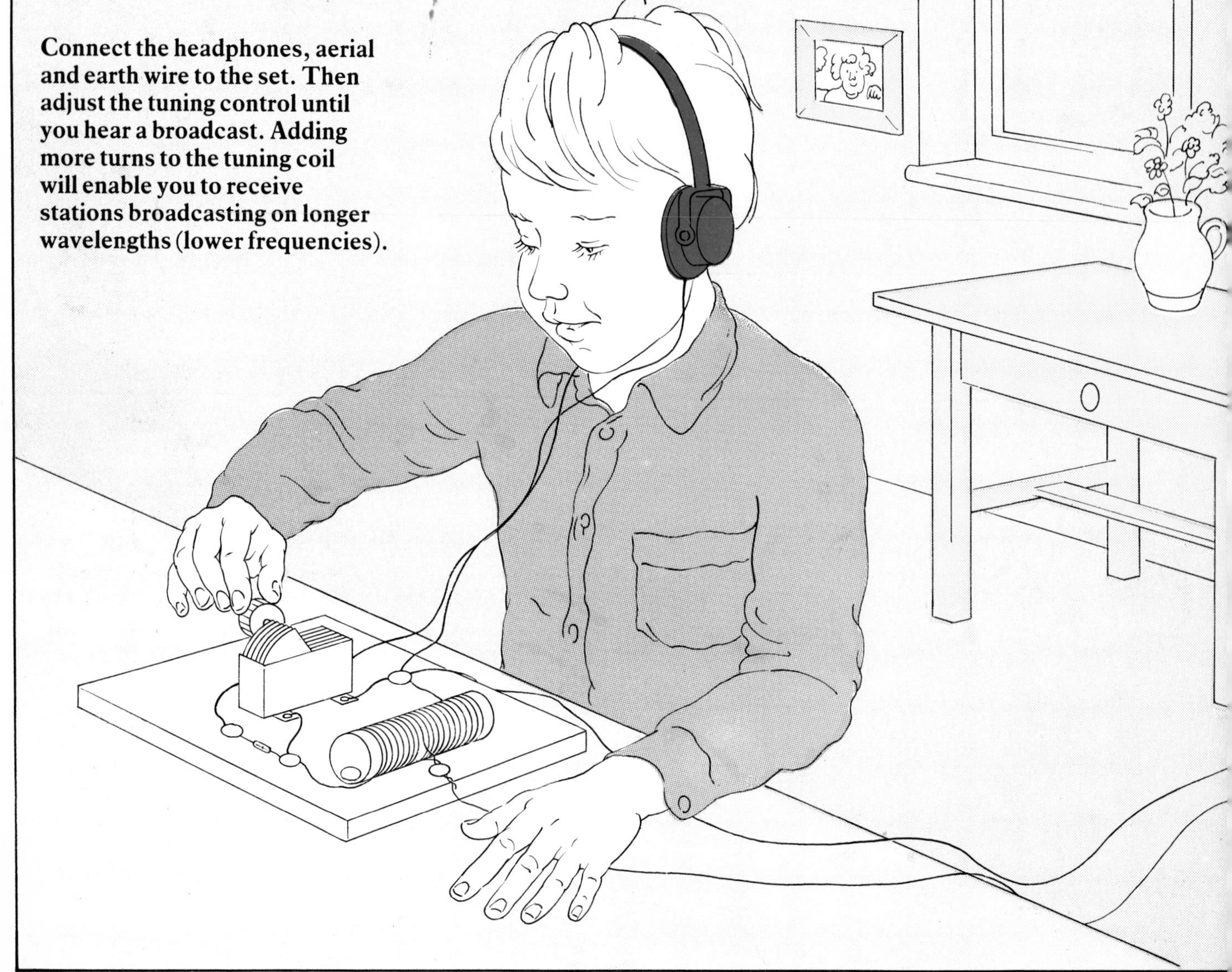

In the first set, you can tune into different radio stations by moving the metal strip along the coil. But the capacitor in the second set gives better tuning control. This set also has an extra connection on the coil for the aerial. Volume is reduced, but stations are easier to separate.

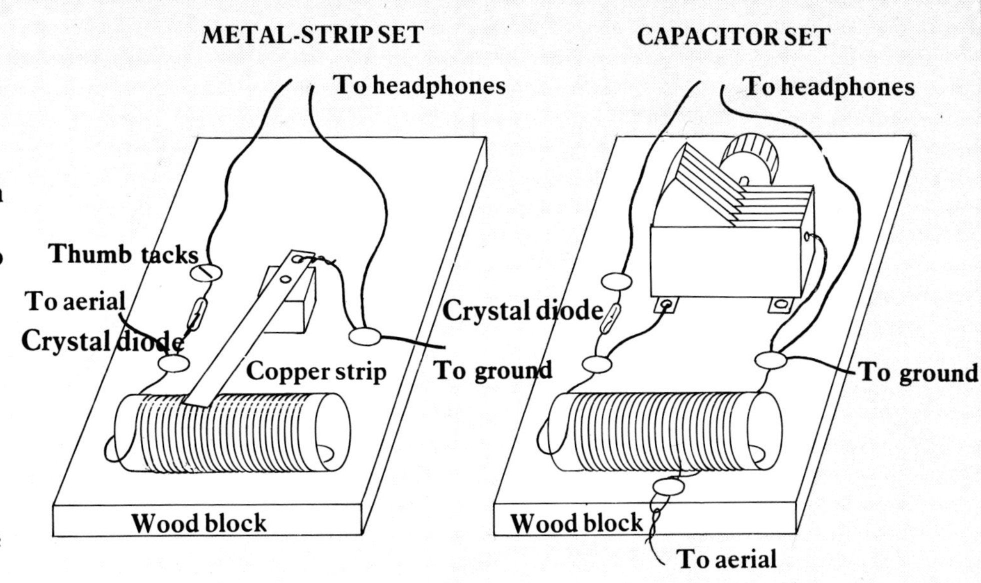

Emery paper

Cardboard tube

Wind the coil around a cardboard tube and tape the ends to prevent slipping. For the first set, use fine emery paper to rub the enamel from the top of the coil.

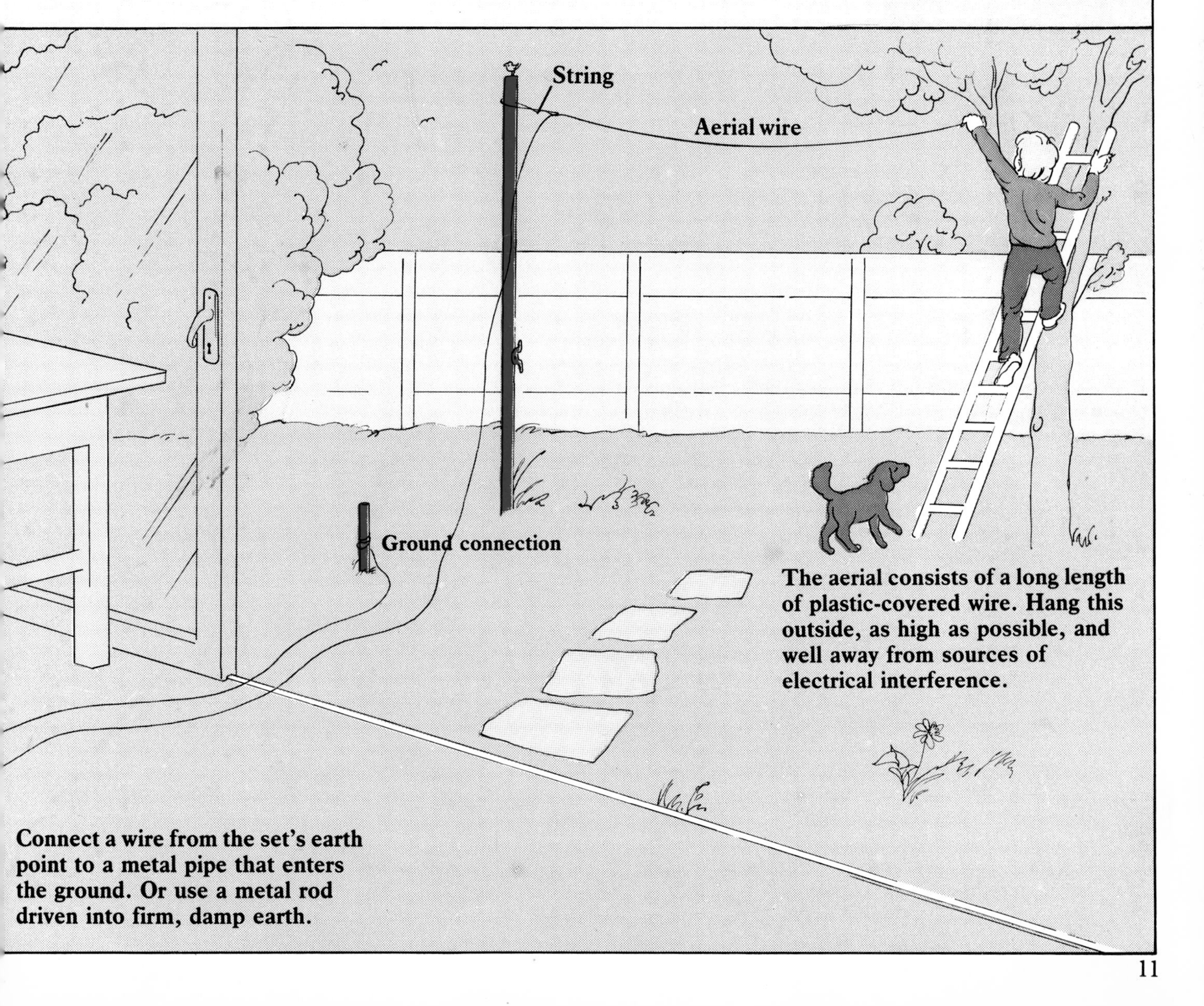

The aerial consists of a long length of plastic-covered wire. Hang this outside, as high as possible, and well away from sources of electrical interference.

Connect a wire from the set's earth point to a metal pipe that enters the ground. Or use a metal rod driven into firm, damp earth.

AMATEUR DETECTIVE

Many criminals are now behind bars because of evidence supplied by police scientists. One important way in which criminals can be identified is from their fingerprints, as no two people have the same fingerprint patterns. Natural oils on our skin are transferred to any object we touch, which is why our fingerprints are usually left behind. Faint prints can be made clearer by applying a fine powder.

The first project shows you how to record fingerprints on paper. Make a collection of several friends' fingerprints, and write each person's name beside his or her marks. Then get one person to make another mark when you are not looking. Now see if you can identify who made the mark.

FINGERPRINTS

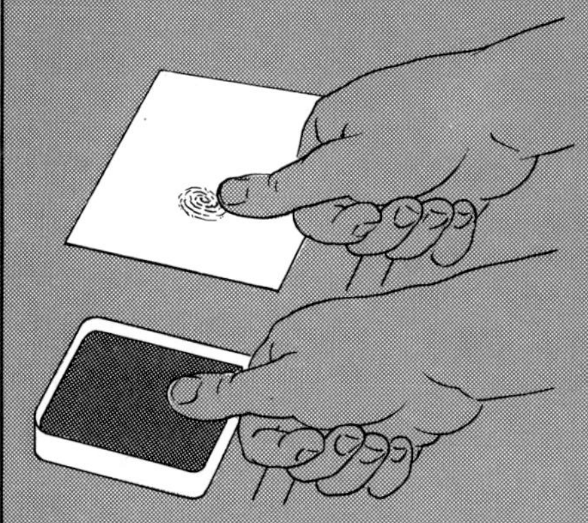

To record a fingerprint, first rub the tip of a finger over a black inking-pad. Then make the print by carefully rolling the finger over a piece of paper.

Plain whorl

Tented arch

Loop

Accidental

All fingerprints are different, but they can be sorted into types, such as those shown above.

PLASTER CASTS

Besides fingerprints, criminals may leave other tell-tale marks behind at the scene of a crime. Footprints or tire marks often lead to the identification and arrest of the offender.

Pour a little water into a bowl. Then stir in some plaster of Paris until you have a thick, creamy mixture. Make sure that there are no lumps in the mixture.

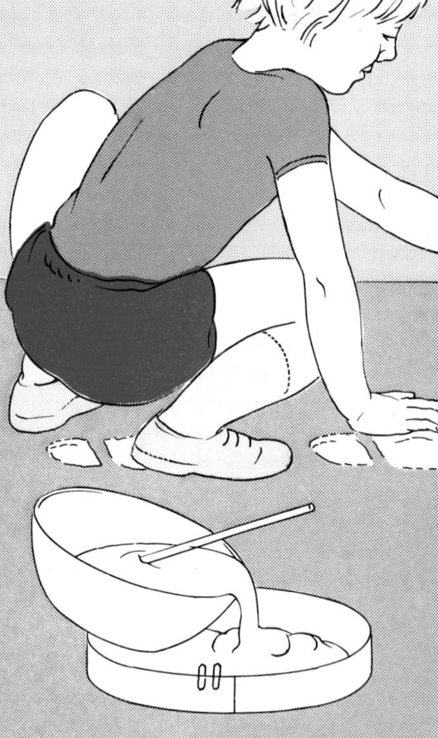

Place a ring of cardboard over the footprint or tire mark to be cast. Remove any loose surface material and then pour the plaster into the enclosure.

The plaster will soon harden. When this has happened, remove the casting from the cardboard. Finally, scrub the casting to remove any dirt.

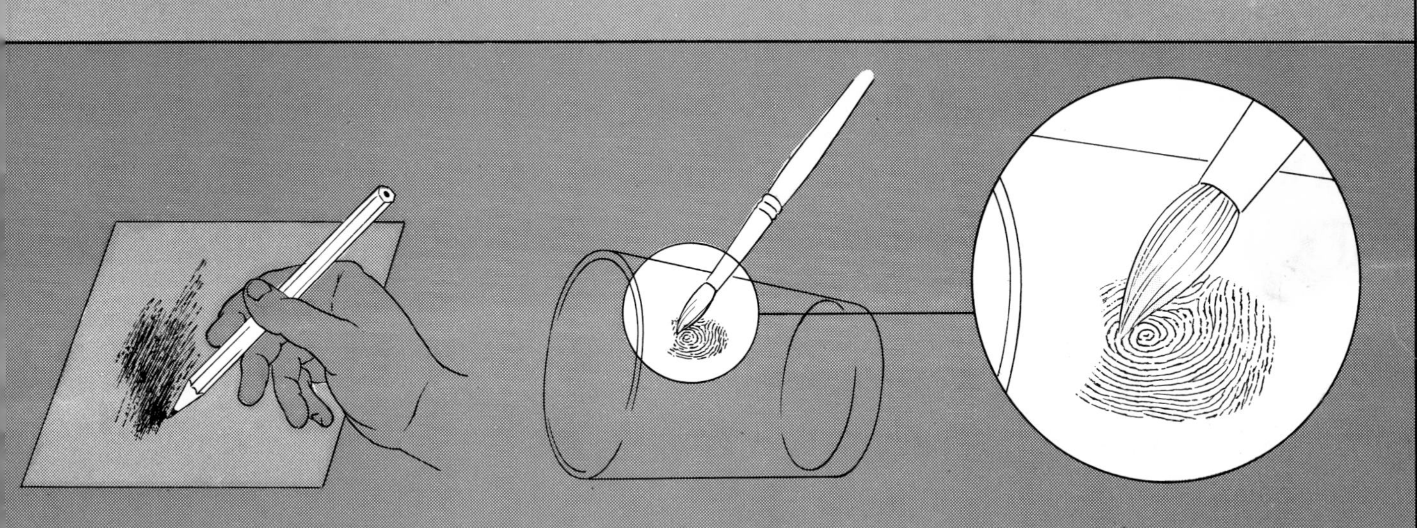

Fingerprints on objects can be made to show up using graphite dust. This can be obtained by rubbing a soft black pencil on a piece of fine sandpaper.

Find a fingerprint on a glass or other smooth surface. Then, using a small, soft paintbrush, carefully apply the graphite dust to the fingerprint.

As you rub the dust over the print, some of the particles will adhere to the oily lines left by the finger. This will make the print much easier to see.

BURGLAR ALARM

A burglar who sets off an alarm will usually run away. And many burglars will not even attempt to enter a place they know to be fitted with an alarm system.

In the alarm shown here, a fine, black thread is stretched across a window or door. Anyone walking into the thread triggers a switch, which makes the alarm bell ring.

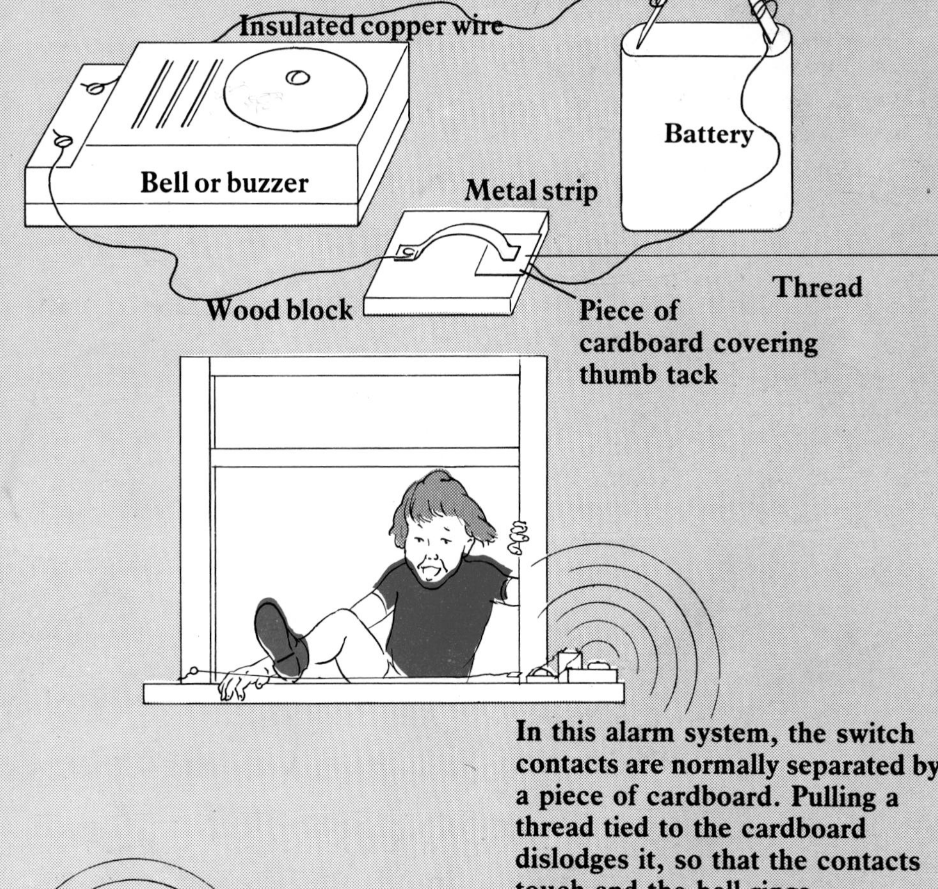

In this alarm system, the switch contacts are normally separated by a piece of cardboard. Pulling a thread tied to the cardboard dislodges it, so that the contacts touch and the bell rings.
The thread can be stretched across a door or window to warn of intruders.

CODES AND SECRET MESSAGES

In the 1830s, an American called Samuel Morse invented the Morse code. This provided a simple way of sending messages by wire, years before the telephone was invented. The code consists of short and long signals called dots and dashes. Letters and numbers are made up from various combinations of these signals.

Messages in Morse can be transmitted in many ways. Often, they are sent as pulses of electricity along a wire or as bursts of radio waves. The signals received may be used to produce sounds. This enables a person knowing the code to translate the dots and dashes back into words.

Details are given here for a Morse code practice set. Two such sets can be used for communication by wire.

MORSE CODE

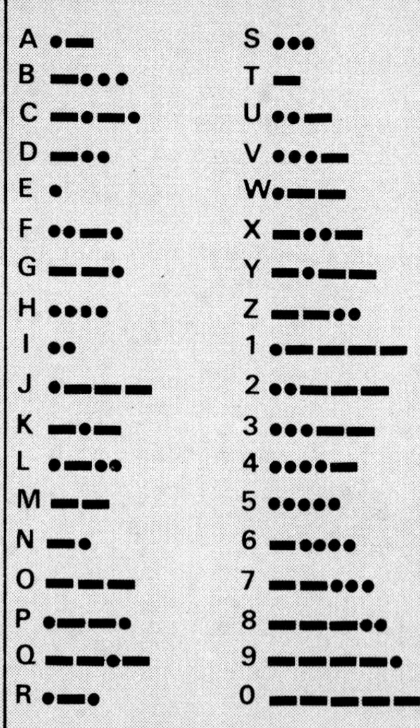

A	•—	S	•••
B	—•••	T	—
C	—•—•	U	••—
D	—••	V	•••—
E	•	W	•——
F	••—•	X	—••—
G	——•	Y	—•——
H	••••	Z	——••
I	••	1	•————
J	•———	2	••———
K	—•—	3	•••——
L	•—••	4	••••—
M	——	5	•••••
N	—•	6	—••••
O	———	7	——•••
P	•——•	8	———••
Q	——•—	9	————•
R	•—•	0	—————

Learn the Morse code using this practice set. The switch, known as a Morse key, is pressed briefly for a dot, and three times as long for a dash. For two-way communication, use two sets with their buzzers connected by a length of twin flex.

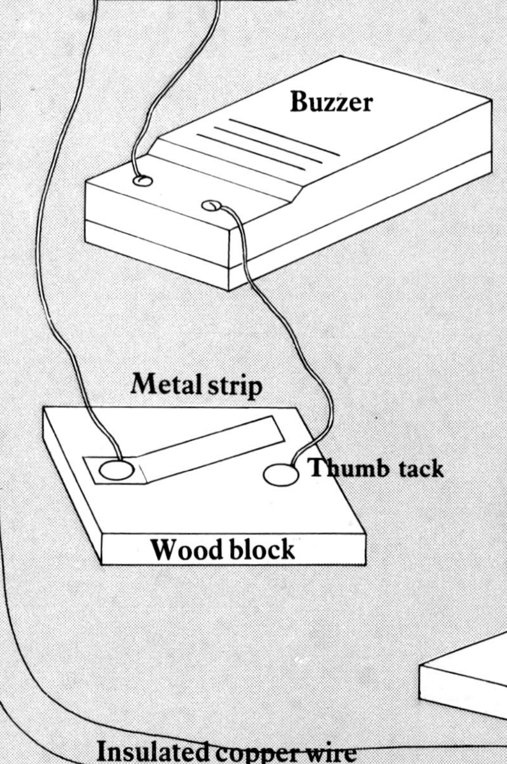

SECRET MESSAGES

Thousands of ways have been devised for sending secret messages. You can change each letter in a message for another, according to a simple code. Or you can jumble up the original letters in various ways. Alternatively, the message can be written in invisible ink or concealed by writing it inside an eggshell.

Put tracing-paper over a page of a printed book. Then, starting at the top, underline letters or words to make up your message. Also trace the page number.

The message can be deciphered (worked out) by a friend with the same book. She has only to put the paper over the correct page in order to read the message.

Messages can be jumbled using squared paper. Write a message across the paper. Then copy the columns of letters formed, as shown here.

The coded message can be read by a friend once he has put the letters back into columns. Be sure to copy down any dashes when doing the decoding.

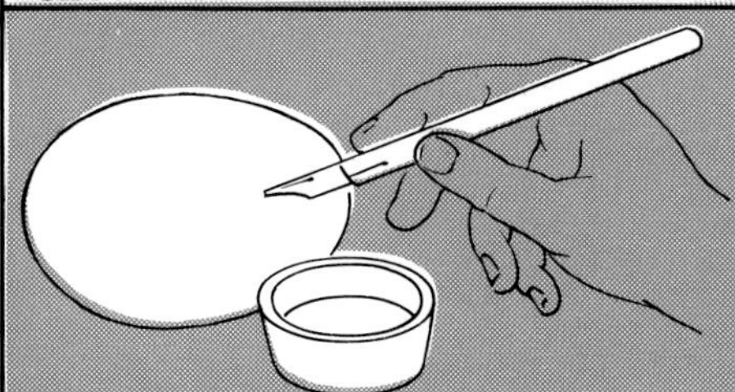

To conceal a message inside a hard-boiled egg, first dissolve some alum (from a drugstore or grocery store) in vinegar. Use this solution to write on the eggshell.

After drying, the message will not be visible. But, by the next day, the solution will have passed through the shell and marked the egg white with your message.

INVISIBLE INK

Invisible ink provides the fastest way of writing a secret message, for the words do not have to be coded in any way. The ink is colorless, but the words can be made visible later by warming the paper.

No special chemicals are required to make invisible ink. Try milk or the juice from a lemon or onion.

Dip a pen or a pointed matchstick into the ink and write your message. Write on blank paper or between the lines of an ordinary message.

Reveal the message by warming the paper gently over a candle flame. The heat causes chemical changes in the ink, which acquires a distinct color – usually brown.

A WEATHER STATION

By weather, we mean the state of the atmosphere in our region. We can make various measurements in order to record the weather and forecast how it will change.

Winds are caused by air moving from high-pressure to low-pressure regions. Changes in local air pressure will therefore cause changes in the winds, and these may bring a spell of different weather. In the northern hemisphere, northerly winds from the polar regions usually bring cooler, drier weather. Southerly winds bring warmer, wetter weather.

The projects here will enable you to check air temperature, wind speed and direction, humidity and rainfall. Humidity means the amount of moisture present in the air.

WEATHER VANE

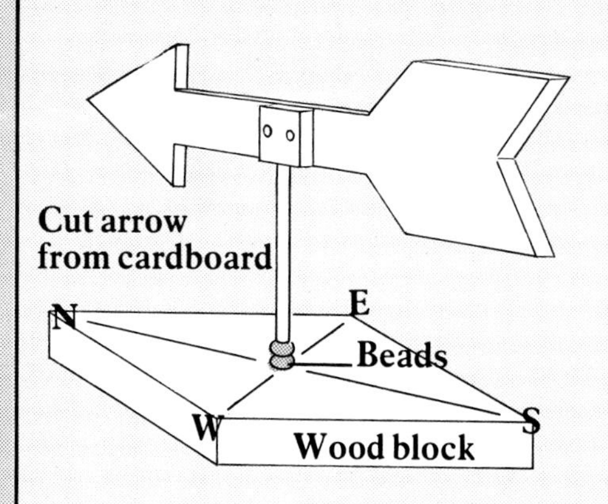

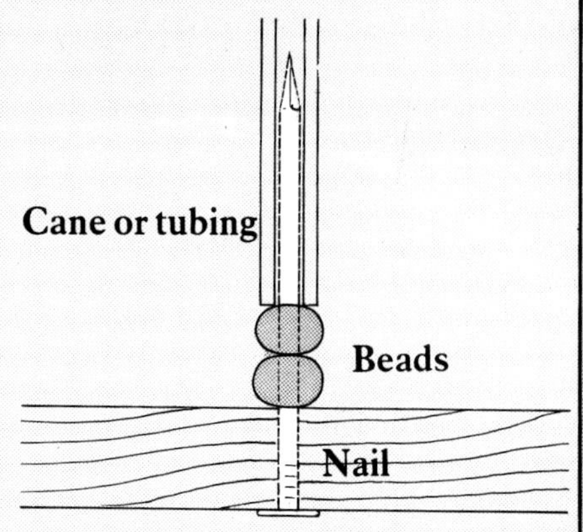

Wind direction can be found using this simple weather vane. The large, cardboard arrow turns freely in the wind to show the direction from which it blows. Mark the points of the compass on the base of the weather vane, then position it so that the north mark points to the north. Check this with a compass.

TEMPERATURE

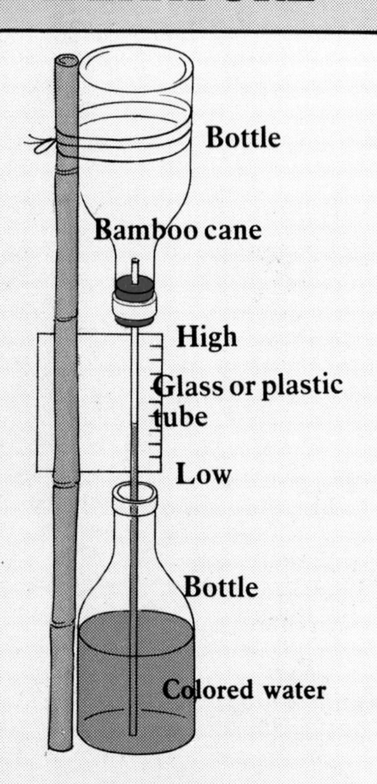

Set up the air thermometer as shown and warm the top bottle with your hands. The air in it will expand and start to bubble out through the colored water. On cooling, water will be sucked up the tube. Temperature changes will alter the water level.

AIR MOISTURE

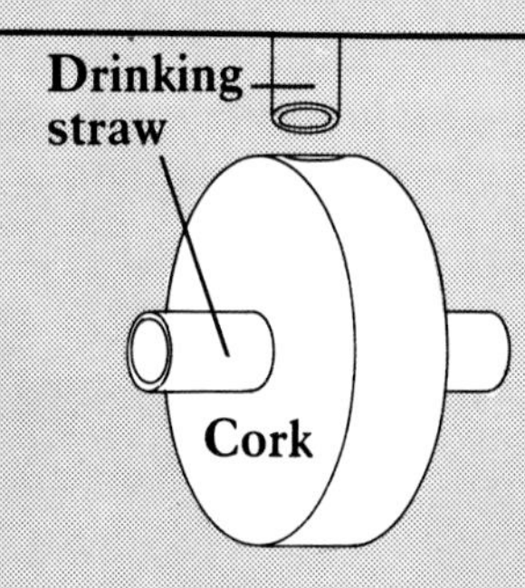

1: Cut a disc from a cork, about the size of a quarter, but ¼ inch (0.5 cm) thick. Make a small hole in the edge and another through the center, then wedge in pieces of a drinking straw.

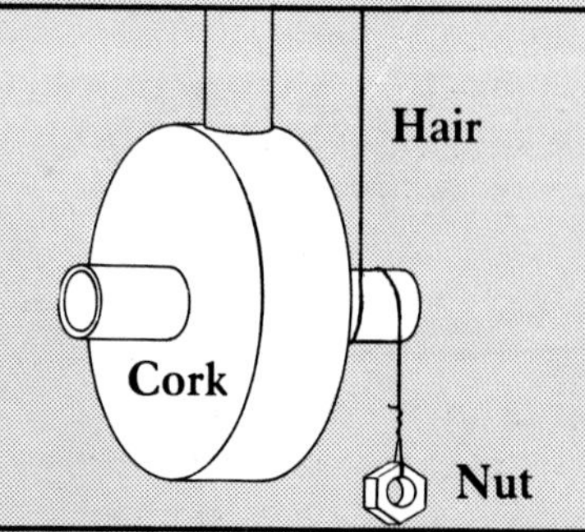

2: Take a long hair and wind it twice around the central straw. Then tie a small weight, such as a nut or washer, to one end of the hair.

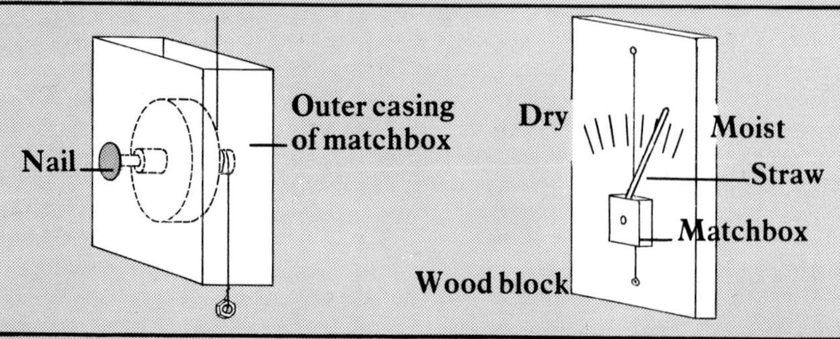

3: Remove the tray from a matchbox and use the outside to house the cork. This must be free to turn on a nail that passes through the box and central straw.

4: Complete construction as shown. In damp conditions, the hair will stretch. Its movement will turn the cork, and the pointer will show high humidity.

WIND SPEEDS

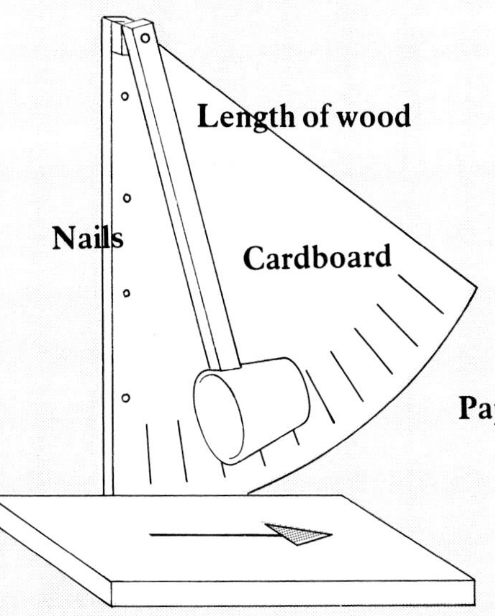

This indicator will enable you to compare wind speeds. It must be turned so that the paper cup faces the wind. The cup will catch the wind and move the arm across the wind-speed scale.

The paper cup is fixed to the wooden arm with thumb tacks. Use glue too for extra strength.

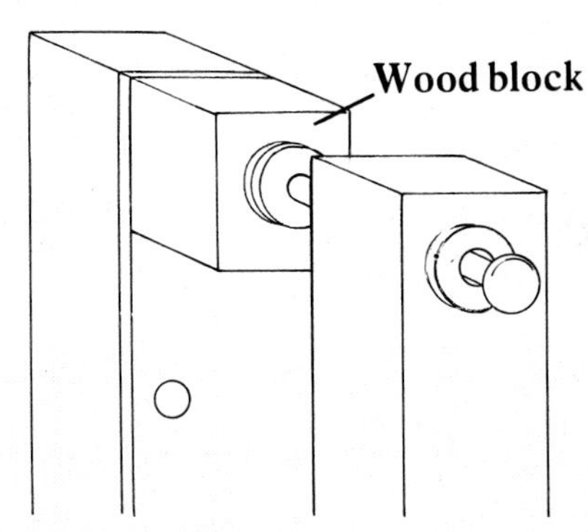

The top of the arm must be drilled through so that it can swing freely on the nail. Use some washers on each side of the arm to prevent it from sticking.

RECORD-KEEPING

To complete your weather station, make a rain gauge, arranged as below. The bottle should have a flat bottom and parallel sides. Trim the top of the plastic funnel to the same diameter as the bottle. Glue the paper strip to the bottle, and mark a scale on it in centimeters or inches, starting at the bottom.

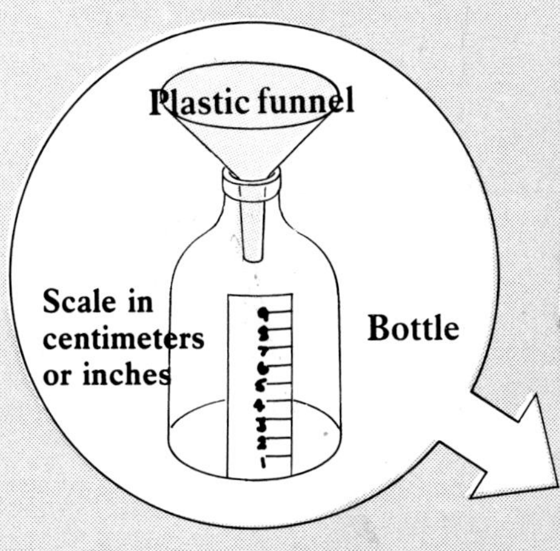

Use your instruments to take readings at the same time each day. Record your results in a notebook so you can see the changes that occur throughout the year.

MICROSCOPES AND TELESCOPES

Most laboratory microscopes contain at least two groups of lenses. But objects can be highly magnified (enlarged) by using only a single lens. This must have a convex (bulging) shape. When light from an object passes through the lens, the rays are bent. This makes the rays coming from a small object appear to be coming from a large object. As a result, we see an enlarged image.

Most lenses are made of glass or plastic, but other transparent substances can be used too. Even water is suitable if it is formed into the right shape. This can be done by using a small, round, wire loop to hold a tiny drop of water. The highly curved surface of the water drop will make it act as a powerful magnifying lens. When an object is placed close to it, an enlarged image will be seen.

WATER-DROP MICROSCOPES

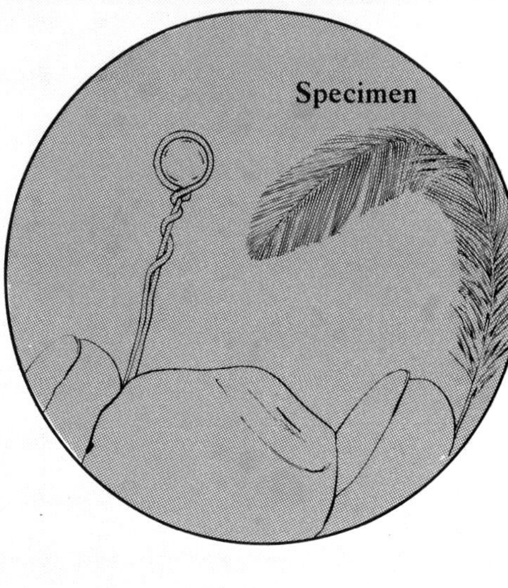

Form a round loop in a piece of wire by winding it around a nail. The loop should be about 1/16 inch (2 mm) across. Use a matchstick to drip a blob of water into the loop.

A COMPOUND MICROSCOPE

This compound microscope has two lenses. At high magnification, it produces sharper images than a single-lens microscope of the same power.

Two convex lenses are required for this compound microscope.

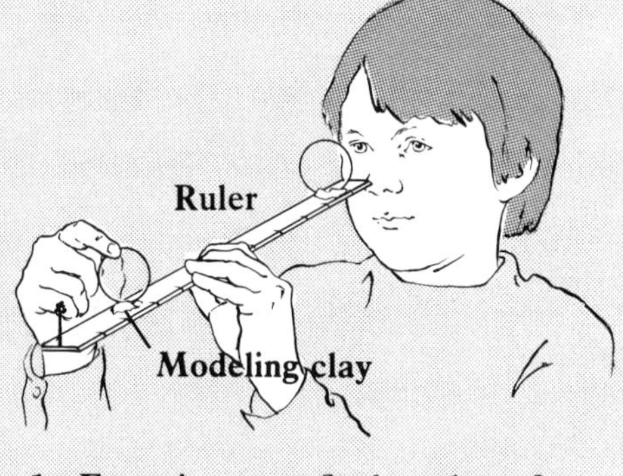

1: Experiment to find out how far apart your lenses need to be. The greater the distance apart, the higher the magnification. Move the object until the image is sharp.

2: When you have determined how far the lenses should be separated, cut a cardboard tube to the same length. Paint the inside with matt black paint.

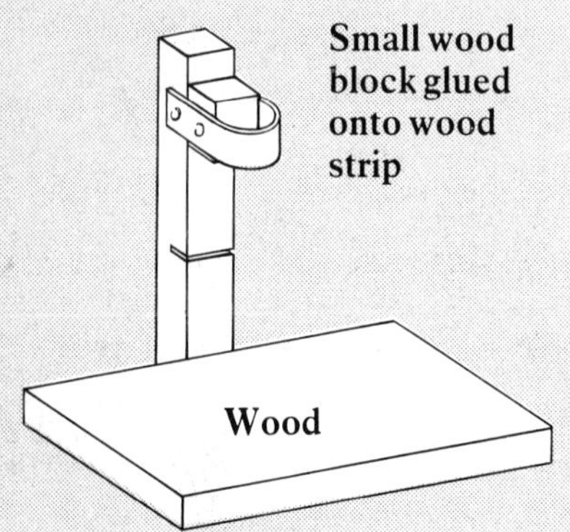

3: Fix the lenses in the ends of the tube. Then make the stand shown above. The loop should grip the tube, but allow it to be moved smoothly for focusing.

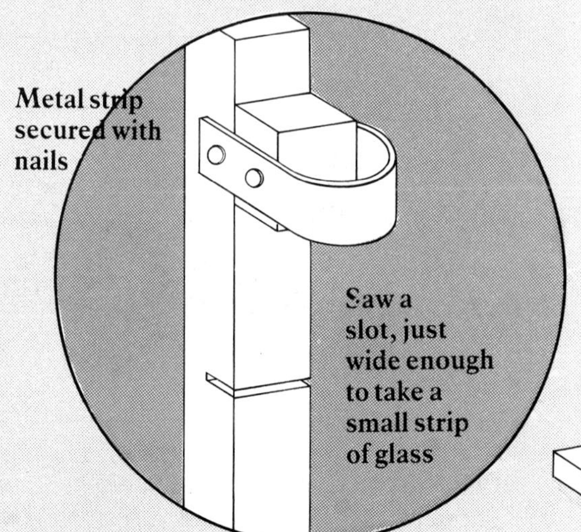

Slide tube up and down to adjust focus

Object being studied

Mirror

Modeling clay

Adjust mirror angle for maximum brightnes

The water-drop microscope can be improved by making a stand to hold the drop at a fixed distance above the object being studied. Transparent objects are best illuminated by using a mirror to reflect through them, as shown.

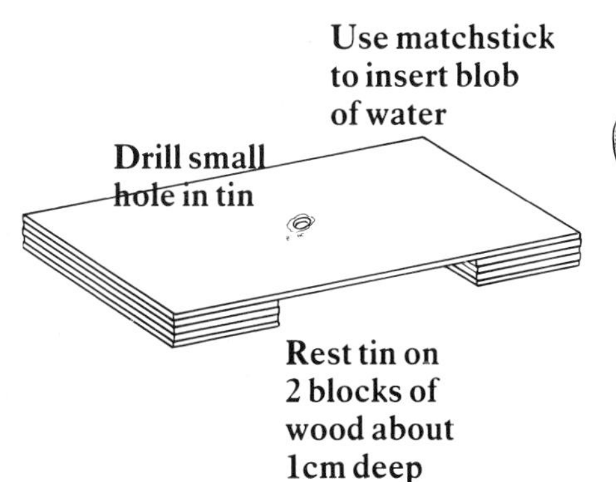

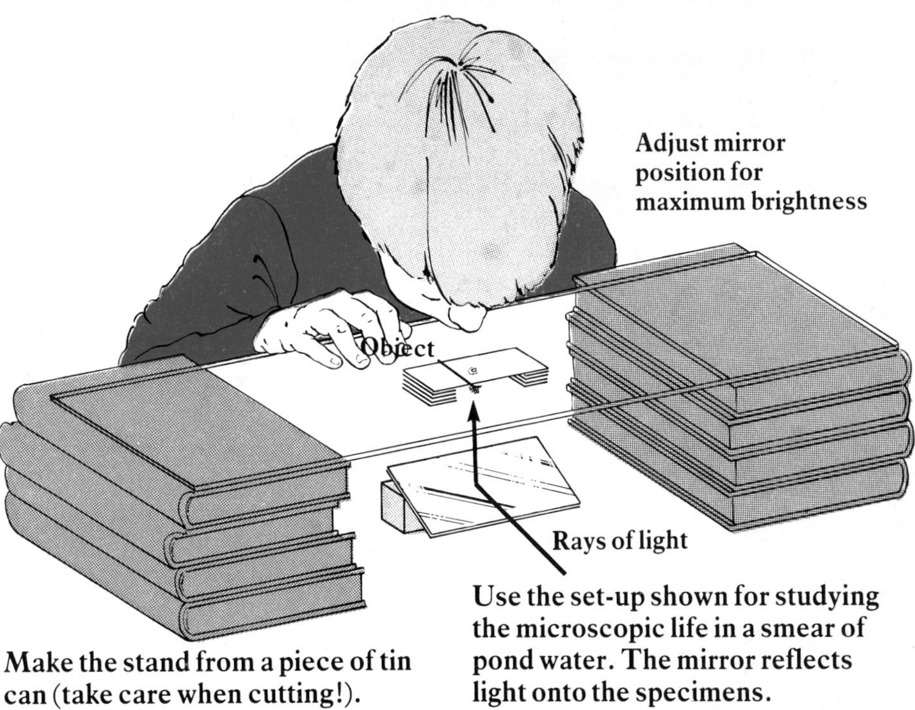

Make the stand from a piece of tin can (take care when cutting!).

Use the set-up shown for studying the microscopic life in a smear of pond water. The mirror reflects light onto the specimens.

A TELESCOPE

The simple telescope uses two lenses to form magnified images of distant objects. If both lenses are convex, the images produced are upside down. For upright images, the lens that forms the eyepiece must be concave (hollowed out).

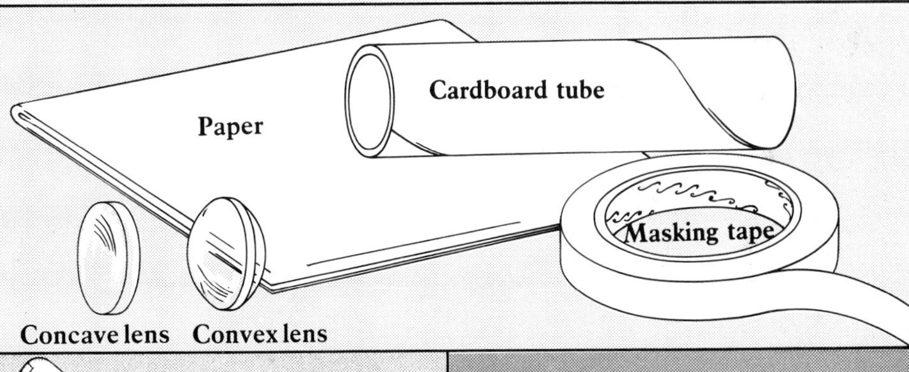

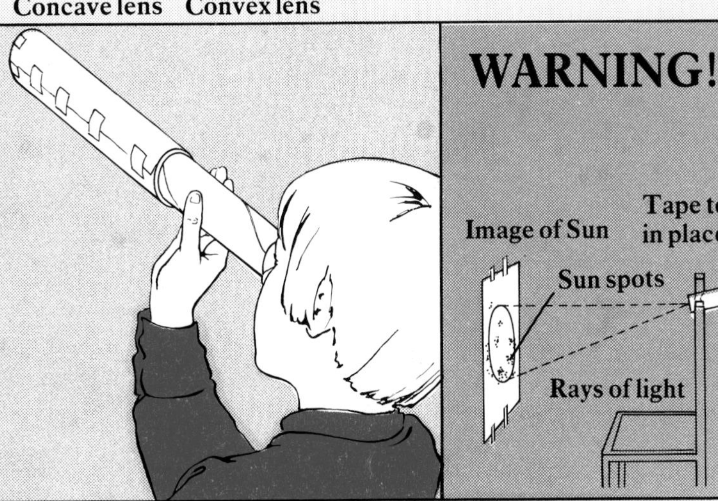

WARNING!

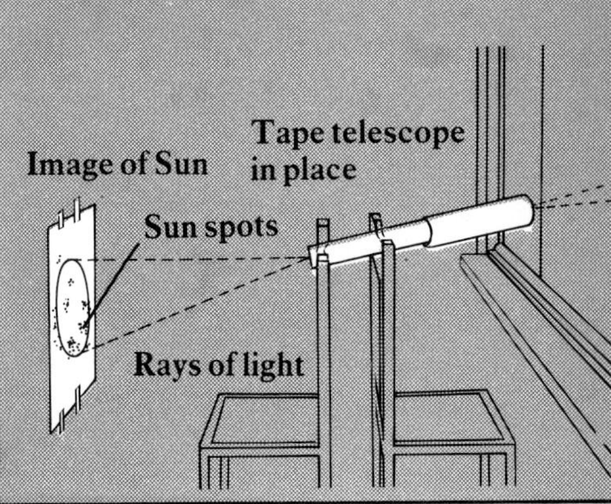

Looking through the concave lens, move the convex lens in front of it. You should soon find a position where you get magnified images of distant objects. Note the distance between the lenses.

You will need two cardboard tubes, each about two-thirds of the distance separating the lenses. One tube should fit tightly inside the other. Or make the outer tube by winding paper around the inner one.

You can use the telescope to observe the Moon, but never look directly at the Sun, as this will harm your eyes. Instead, use the telescope to project the Sun's image onto a piece of paper.

REFLECTIONS

Interesting optical effects can be produced using two small mirrors. A single mirror forms just one image of an object. But two mirrors placed at an angle to each 'other can produce numerous images. This effect is used in a device called a kaleidoscope. Small pieces of colored paper in the kaleidoscope are reflected by a pair of mirrors. The multiple images produced form beautiful patterns. Shaking the instrument changes the positions of the objects, so that a new pattern is formed.

A different arrangement of two mirrors forms the basis of a simple periscope. This will enable you to see over a high fence or wall. And, in the last project, a sheet of glass creates the illusion of a ghost.

MIRROR IMAGES

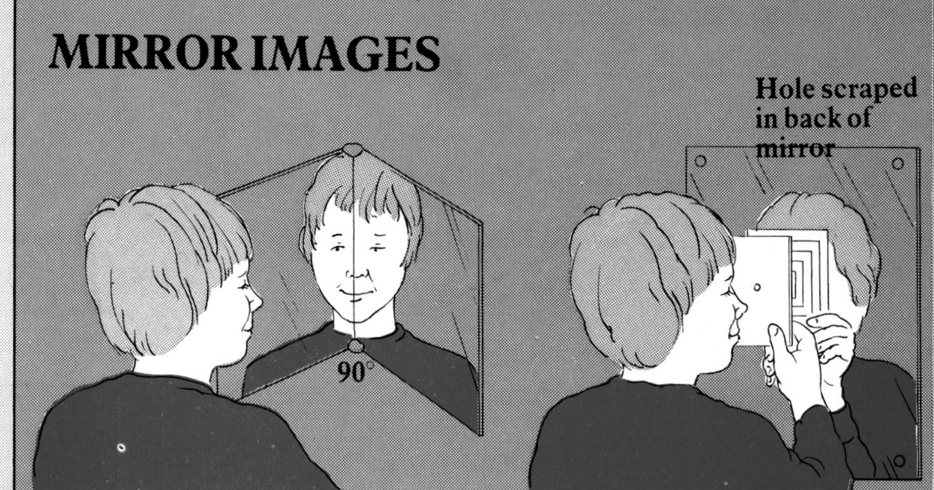

In an ordinary mirror, our image is reversed. But two mirrors at right-angles give a right-way-round image. (See how confusing this is when combing your hair.)

If two mirrors face directly towards each other, light can bounce back and forth between them. Peer through one mirror and see the unending series of images.

KALEIDOSCOPE

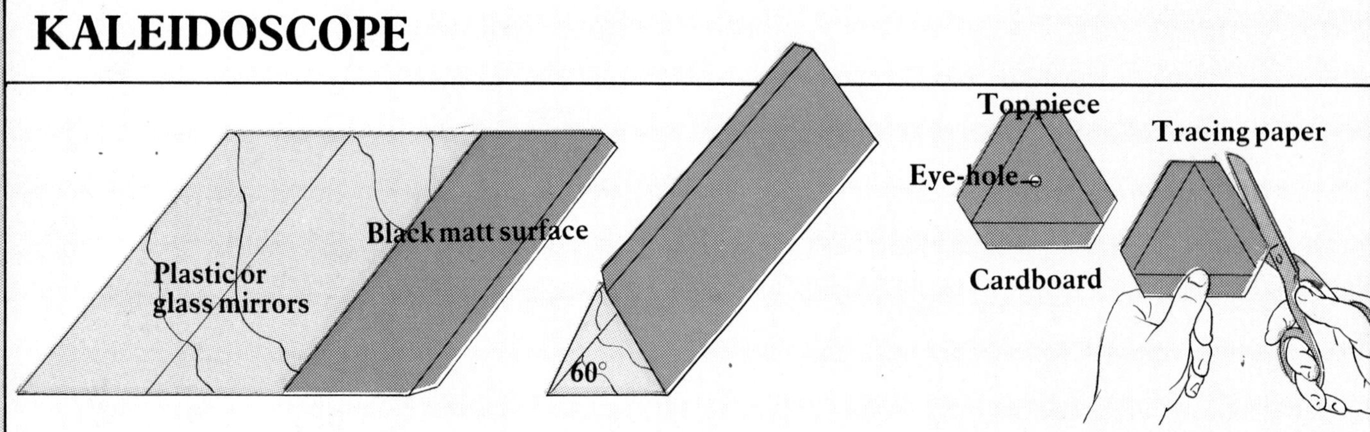

This kaleidoscope consists of two narrow mirrors in a cardboard tube. You can use plastic mirror tiles or an ordinary pocket mirror cut into two.

The mirrors should be identical in size. Make a triangular cardboard tube with sides just larger than the mirrors. Then glue the mirrors inside the tube.

The angle between the two mirror surfaces should be exactly 60°. Make a top piece for the tube out of cardboard, and cut the bottom piece from tracing paper.

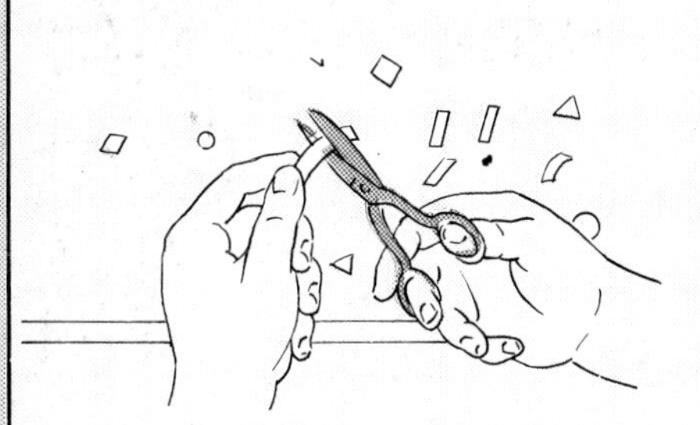

Glue the top piece in place, and then cut up some tiny pieces of colored paper or, better still, cellophane. Use a few pieces of as many different colors as you can find.

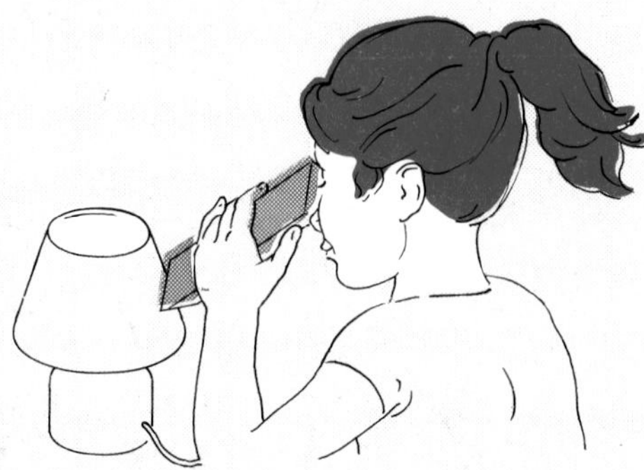

For best results, point the kaleidoscope at a light. Simply shake the tube to move the scraps and change the pattern.

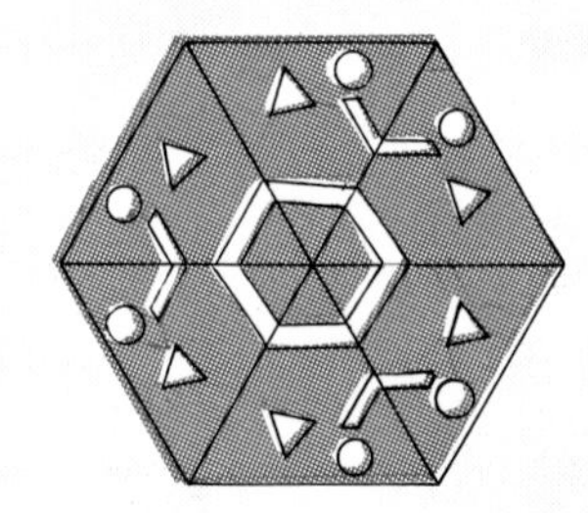

Put the coloured scraps inside the kaleidoscope and then glue on the bottom of the instrument. Interesting patterns will now be seen if you look through the eye-hole in the top.

PERISCOPE

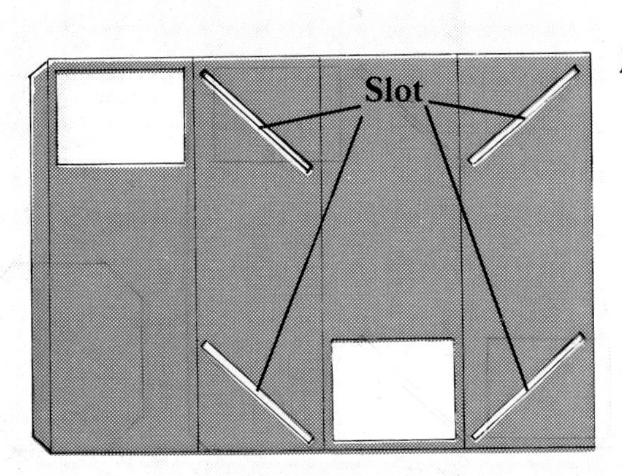

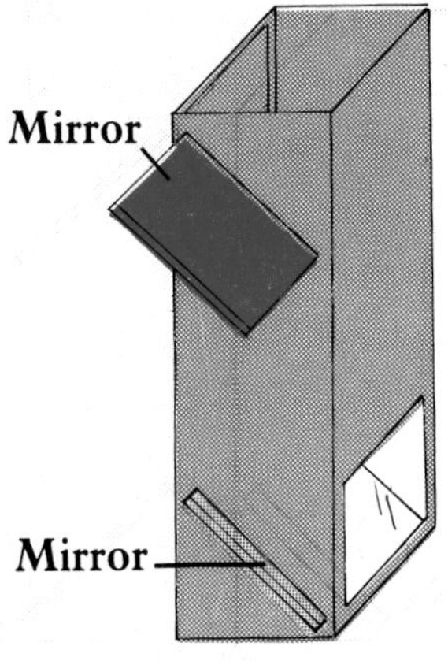

You will need two small mirrors the same size and a sheet of stiff cardboard. On the cardboard mark out four equal panels and cut four slots at 45° to the horizontal, as shown. These are to hold the mirrors in position. Fold the cardboard to form the periscope tube and tape the edges together. Push the mirrors into position.

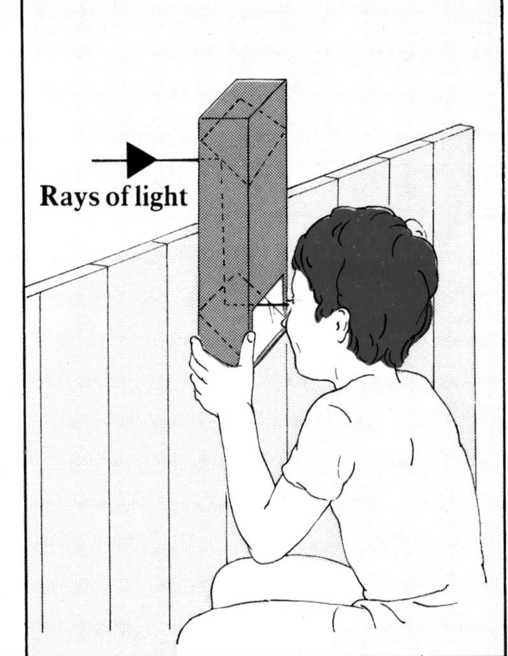

PEPPER'S GHOST

This famous stage trick was popular in Victorian times. The ghost can be made to appear or vanish at will and can appear with other actors or objects. Unknown to the audience, the actor playing the ghost is not on stage. He or she is out of sight at the side. The audience can see the ghost's reflection in a sheet of glass going right across the stage.

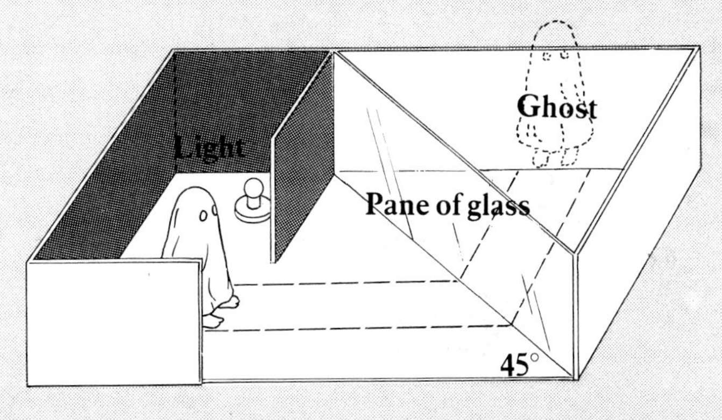

This model stage can be made from wood, cardboard, or a combination of the two.

The glass strip should be fixed at an angle of exactly 45° to the front of the stage.

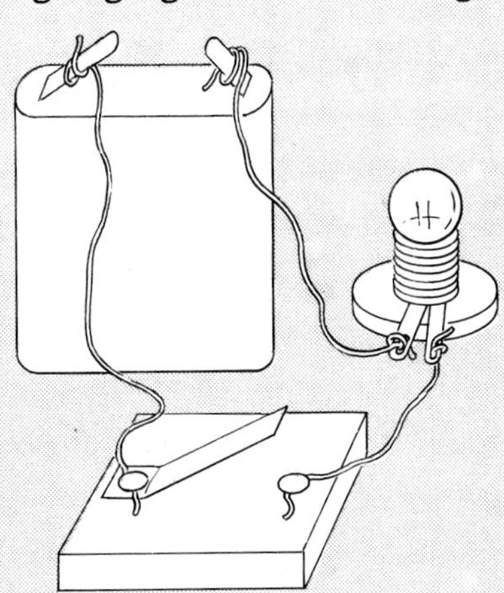

A small lamp is used to illuminate the model ghost. Connect the lamp to a battery and switch, as shown. Pressing the switch will light the lamp.

Curtains can be added at the front to hide the model ghost.

When the light is switched on, a ghostly image appears on stage.

OPTICAL ILLUSIONS

We experience optical illusions almost every day of our lives. When we look into a mirror, the images we see appear to be behind the mirror. But this is an illusion – the images are not really there at all. When we go to the movies, we see what appear to be moving pictures. But we are really watching a series of thousands of still pictures, each slightly different from the one before. The pictures are changed too rapidly for the eye to detect, so we have the illusion of seeing a single, moving picture. The flick book demonstrates this well.

Our eyes can deceive us in many other ways. Separate images may sometimes appear to combine and we may misjudge the length or straightness of a line.

TRICK OF THE EYE

In everyday life, we usually believe what we see. With an illusion, we believe something that we only *think* we see. Can you trust *your* eyes?

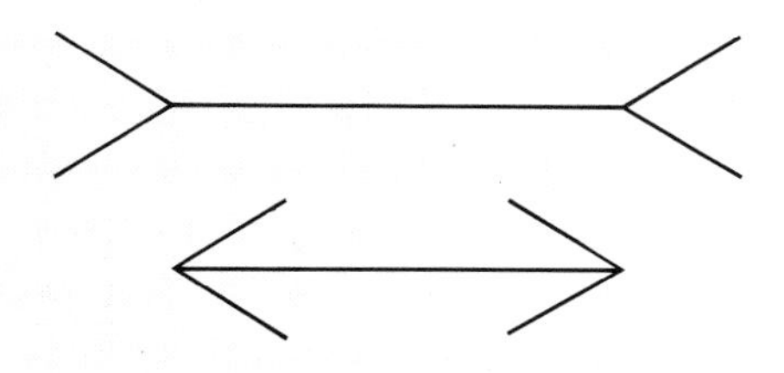

Which of the two horizontal lines above is the longer? Use a ruler to check your answer.

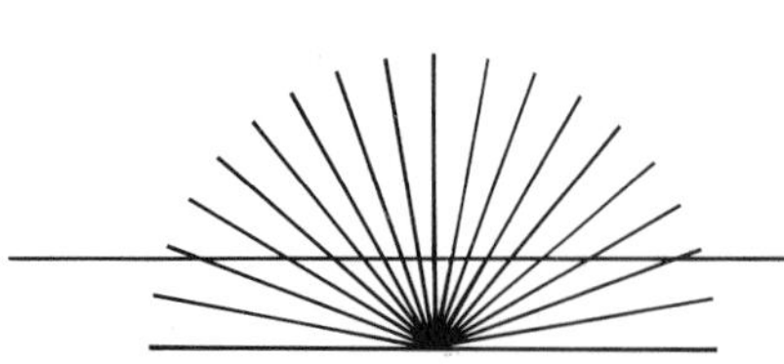

Does the line going through the middle curve? Use a ruler to check your answer.

Stare at the dot with both eyes. Now move the book towards you and see the fish swallow the worm.

FLICK BOOK

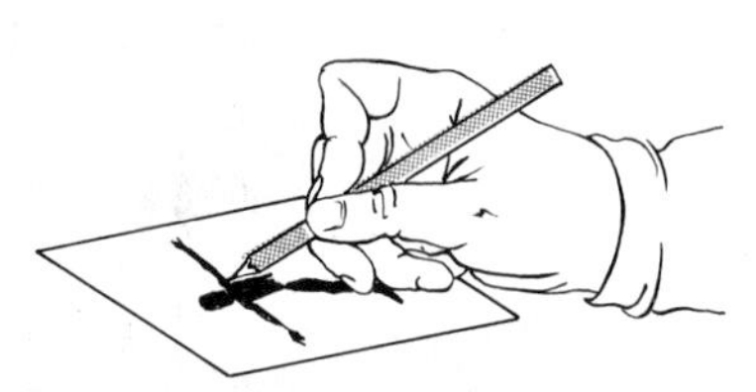

1: Draw 18 pictures of a person jumping, each slightly different from the one before. If you wish, use the pictures below as a basis.

2: Each drawing must be the same size and at exactly the same place on each piece of paper. Use the technique shown above to do this.

3: These pictures show six main stages in a sequence. You can work out the stages in between for yourself. Pictures 17 and 18 should show the person returning to position 1. For best results, repeat the series several times.

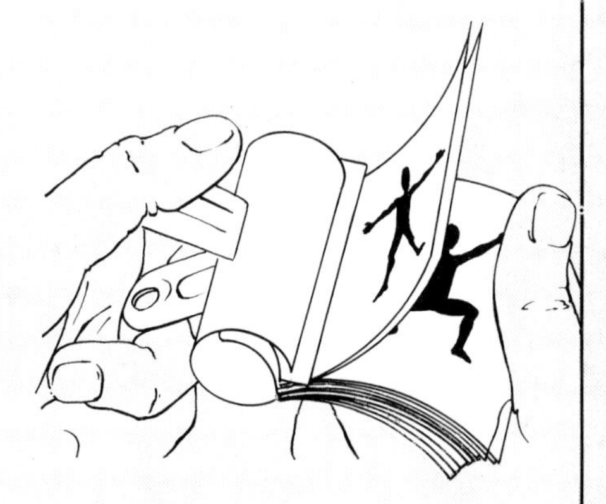

4: Insert the pictures, in order, between the jaws of a bulldog clip. Now flick through the pictures quickly and see the person appear to come to life!

A MOVING FISH

After an object has disappeared from view, our eyes retain its image for a fraction of a second. Here, a new image is formed before the previous one has died away. So the two images appear to merge.

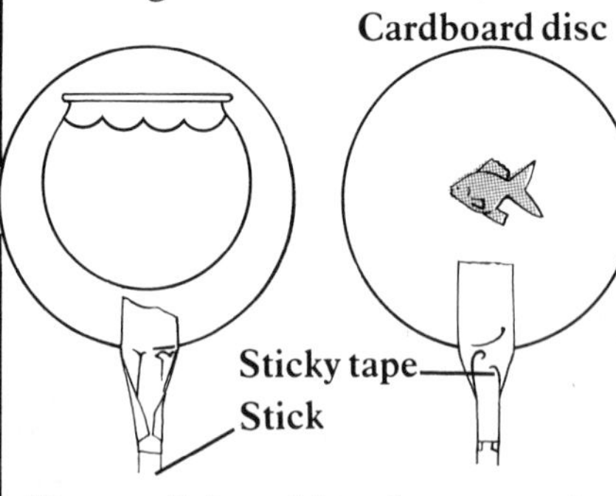

Draw a fish and bowl on opposite sides of a cardboard disc. Fix the card in the end of a stick.

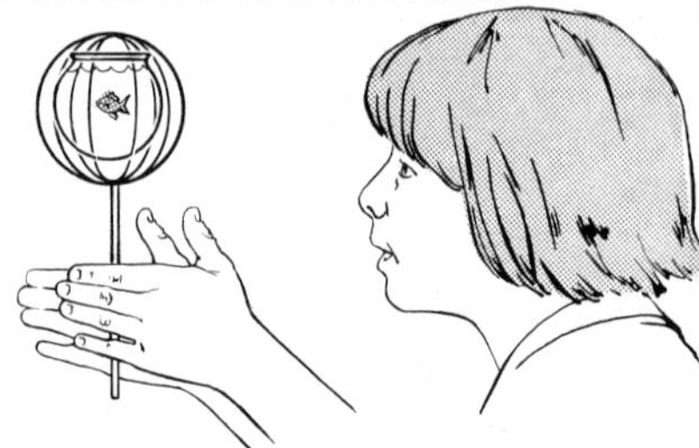

Make the stick turn quickly between your hands. The fish will appear to be in the bowl.

PRAXINOSCOPE

The praxinoscope entertained children with moving pictures long before the first cinemas were opened. A series of slightly different pictures were printed inside a circular band. This was turned quickly, and the pictures viewed through slots from the outside.

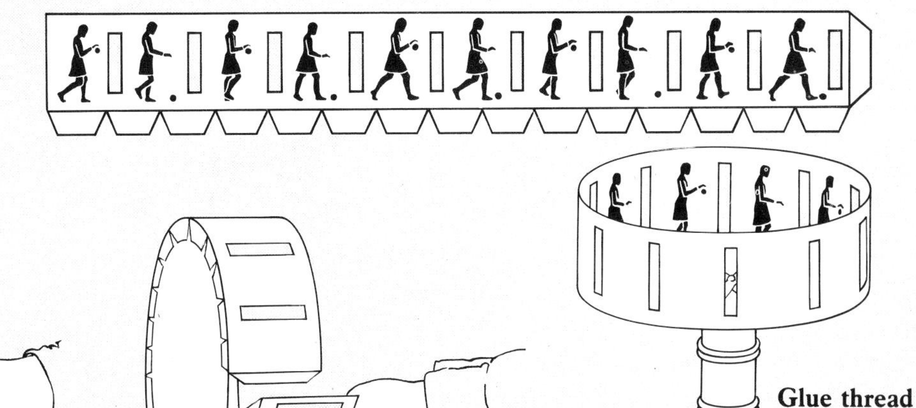

1: Place a large, round tin on a sheet of cardboard and draw around it. Then cut out the circle of card.

2: Cut a strip of thin card to fit around the disc. Make slots and tabs at regular intervals, as shown above.

3: Between the slots, draw a series of pictures showing a person moving. Complete construction by pushing a nail through the center of the disc into some thread spools.

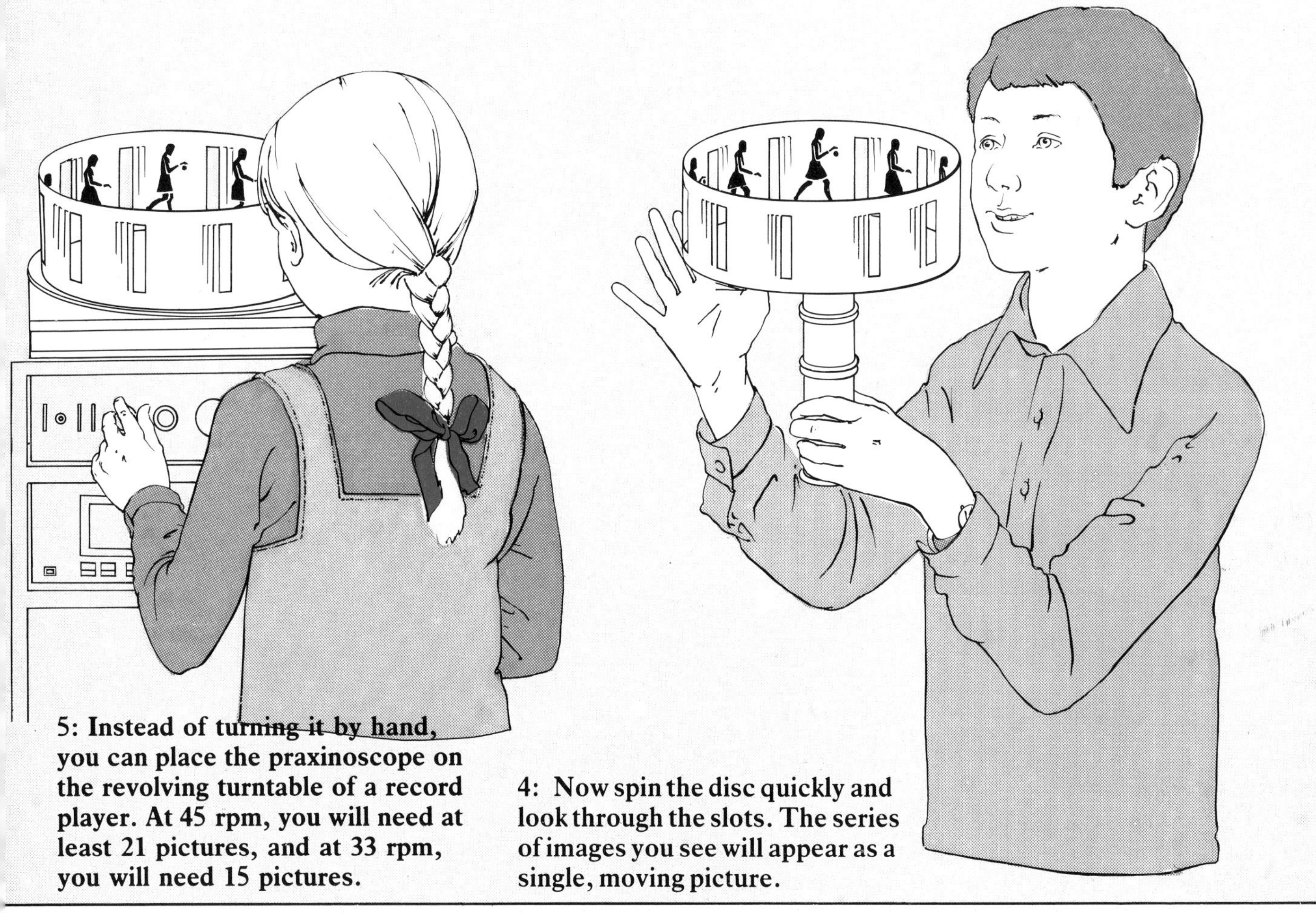

5: Instead of turning it by hand, you can place the praxinoscope on the revolving turntable of a record player. At 45 rpm, you will need at least 21 pictures, and at 33 rpm, you will need 15 pictures.

4: Now spin the disc quickly and look through the slots. The series of images you see will appear as a single, moving picture.

MAKING THINGS MOVE

To make anything move, some force must be applied to it. The small boat on the right is moved by forces of attraction (known as surface tension) in the water surface. Camphor reduces the forces behind the boat so that the forces in front pull the boat forward.

The model paddleboat is propelled by energy stored in a twisted rubber band. As the band unwinds, it turns a paddle, which exerts a force on the water. Whenever a force (action) is applied by an object, that object experiences an equal, but opposite force (reaction). Here, the reaction of the water on the paddle pushes the boat forwards. Like the paddleboat, the rocket boat and propeller-driven car are both moved by forces of reaction.

CAMPHOR BOAT

Construct a boat, as shown below. With a piece of camphor (mothball) inserted in the notch, the surface tension effect will move the boat slowly across a bowl of water. (The water must be absolutely clean and still.)

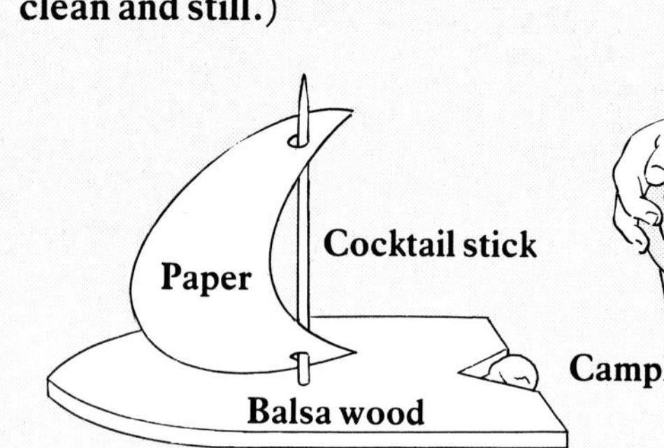

STEAM JET BOAT

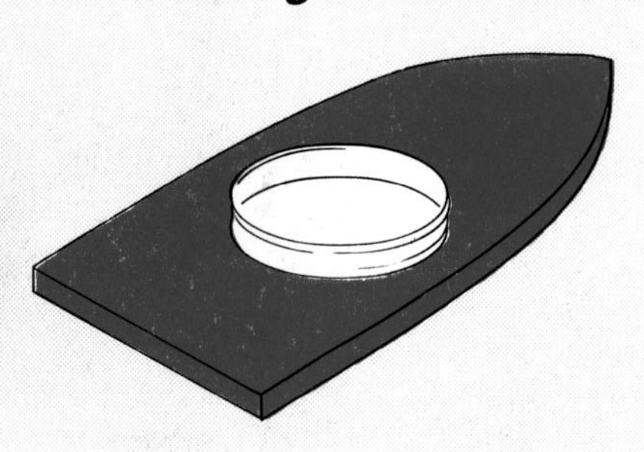

Use balsa wood for the base of the boat. Do not make it too narrow, or it will be liable to capsize. Glue on a jam jar lid to protect the wood from the flame.

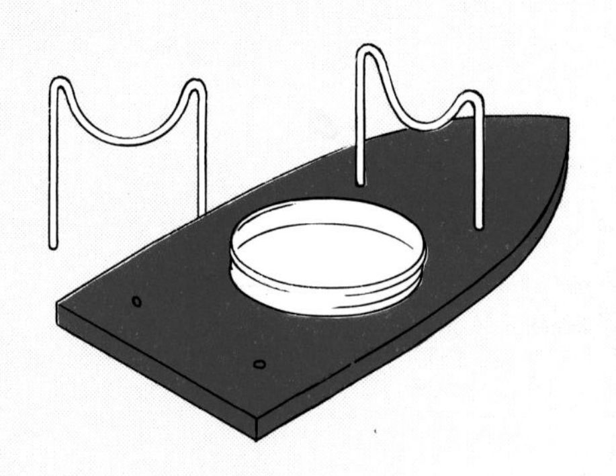

Get two pieces of stiff wire and bend them to shape, as shown. Push the ends of the wires halfway through the wood. These act as supports for the engine.

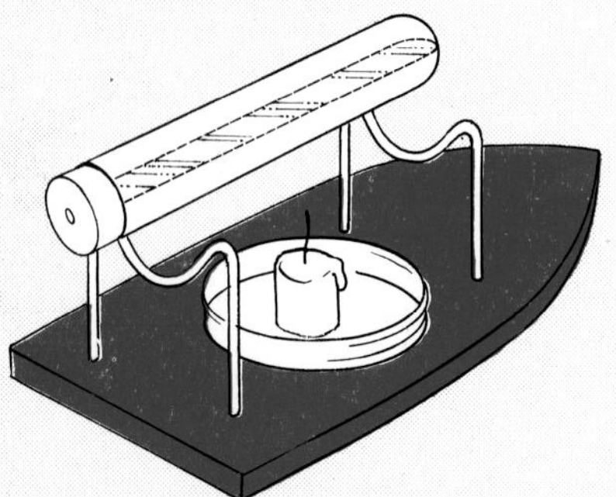

Make a small hole in the cap of a metal cigar tube. Put a little hot water in the tube and then screw on the cap. It is important that this should fit tightly.

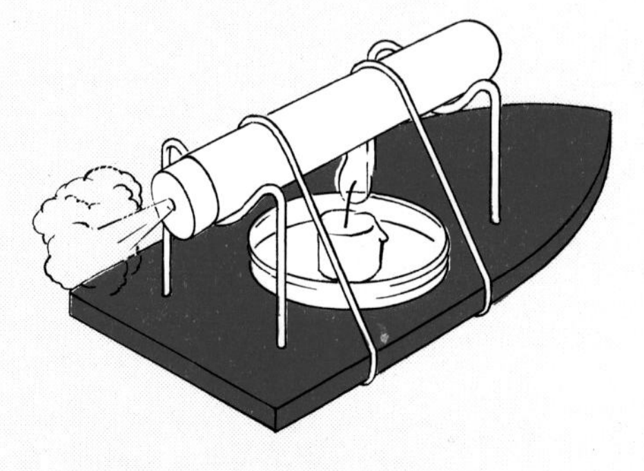

Secure the tube with rubber bands, and light a candle stump placed under the tube. Soon, the water will boil, and the steam jet will propel the boat.

WHIRLING DERVISH

Use scraps from a cork to make the central support and four outer pieces. Fix these together with cocktail sticks or sharpened matchsticks. Insert pieces of camphor in the notches. Cut the dervish figure from cardboard and glue onto the central support. Then place in clean, still water.

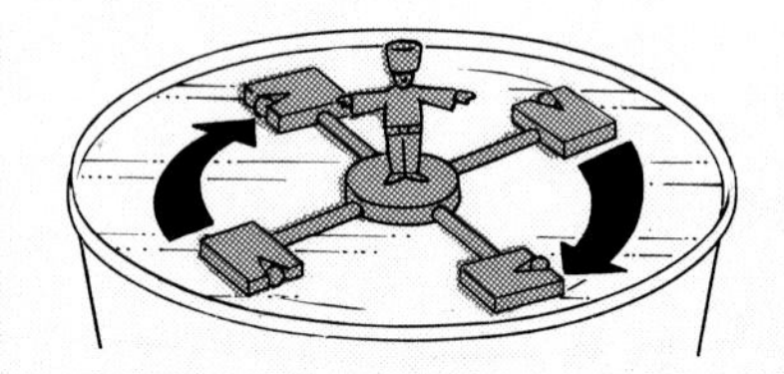

Watch the dervish start to spin, as surface-tension effects pull the outer corks through the water.

RUBBER-BAND PADDLEBOAT

Cut the base of the boat from a piece of balsa wood. For the paddle, use two pieces of thinner wood slotted together. Support the paddle with a rubber band stretched across the slot.

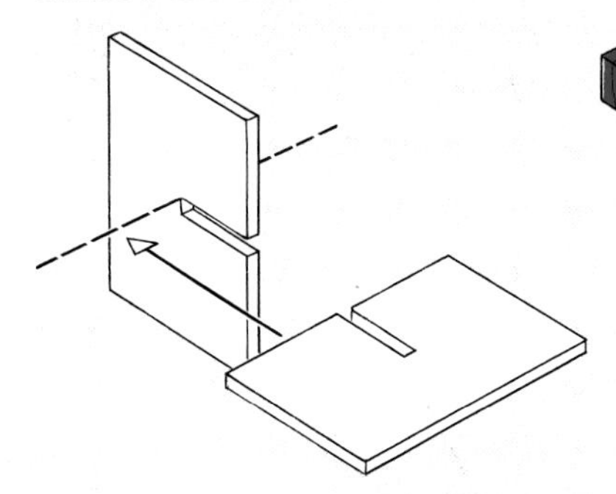

Turn the paddle several times so that the band becomes twisted. Then release the paddle, and see it turn and propel the boat forward.

PROPELLER-DRIVEN CAR

In this car, a small electric motor turns a propeller which drives the car forwards. A model-aircraft propeller can be used, or you can carve your own from balsa wood. Construct it as shown below. Reverse the battery connections if the car goes backwards.

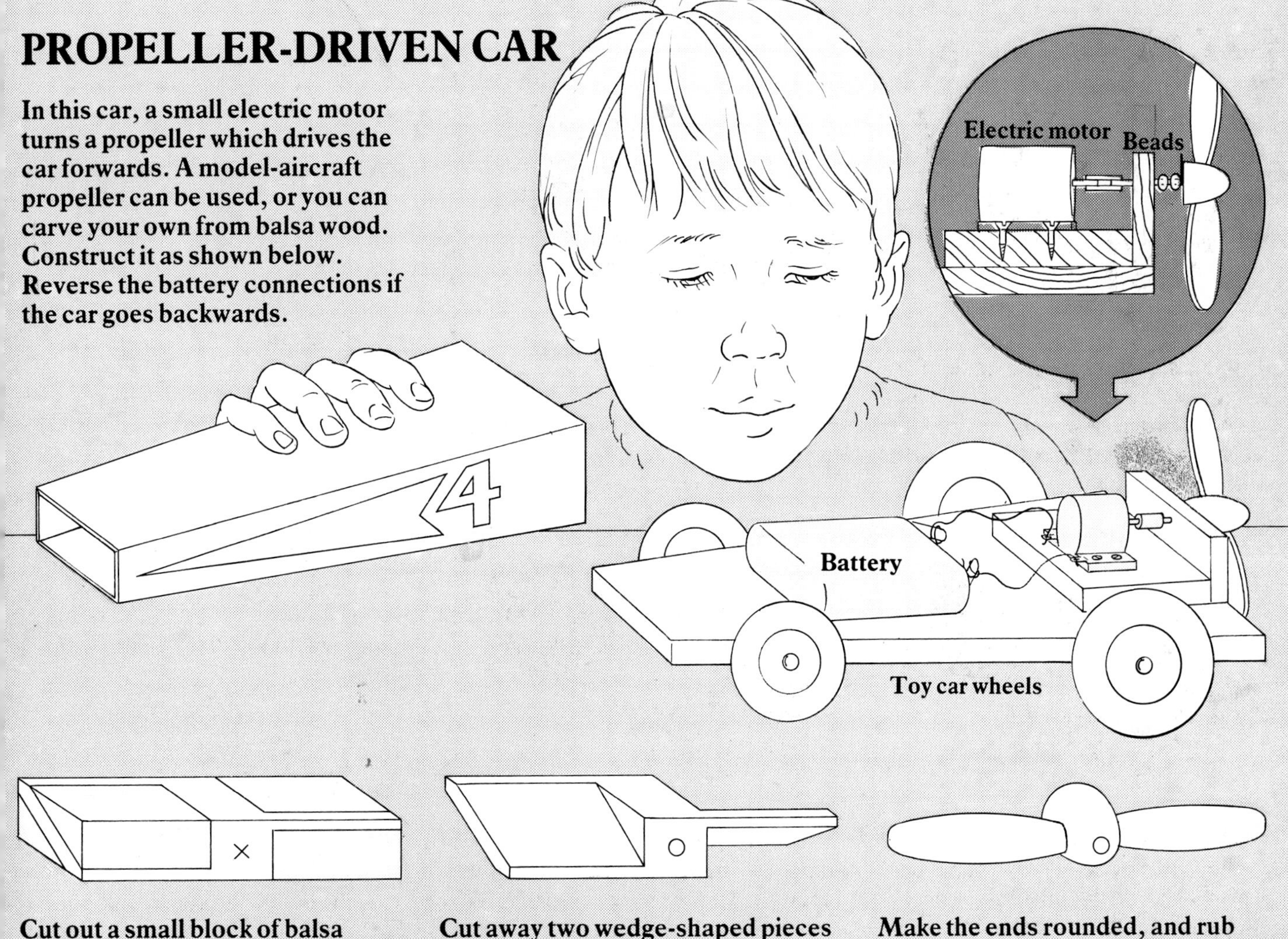

Cut out a small block of balsa wood, and mark as shown above.

Cut away two wedge-shaped pieces from each end of the block.

Make the ends rounded, and rub with sandpaper for final shaping.

PUZZLES AND TRICKS

In the first trick, shown on the right, an effect called surface tension pulls a cotton loop into a near-perfect circle. The hovering card is a good demonstration of what scientists call Bernoulli's principle. This states that the pressure of a gas decreases when it moves quickly. In this case, the gas is air, which is blown over the upper surface of a card. As a result, the pressure above the card decreases, enabling the higher pressure below to hold it up.

The water trick demonstrates that air pressure can support a column of water. And the wood-breaking trick depends on the pressure of air around us. Air pressure is also used in the final trick to force an egg into a bottle.

COTTON LOOP TRICK

Mix some liquid detergent with water to make a strong, soapy solution. Dip a large wire loop in the solution to form a soap film over the wire.

Make a small loop of cotton thread and place it carefully on the soap film. Make sure that all of the cotton loop is in contact with the soap film.

THE HOVERING CARD

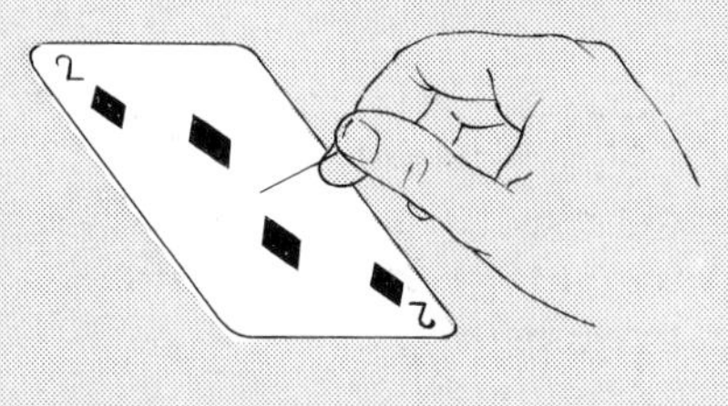

Push a pin through the center of a card. Then place the card against the end of a thread spool so that the pin goes into the hole.

Holding the card in place, start to blow through the hole in the spool. Tilt the spool so that the card is facing the floor.

Release the card while continuing to blow. You may be surprised to find that, no matter how hard you blow, the card will not fall.

THE MAGIC GLASS

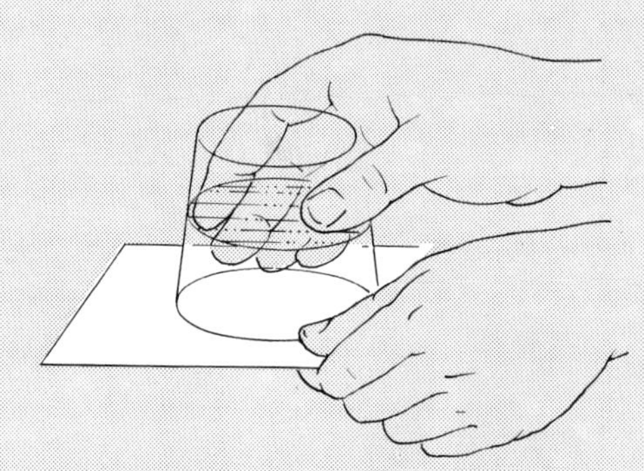

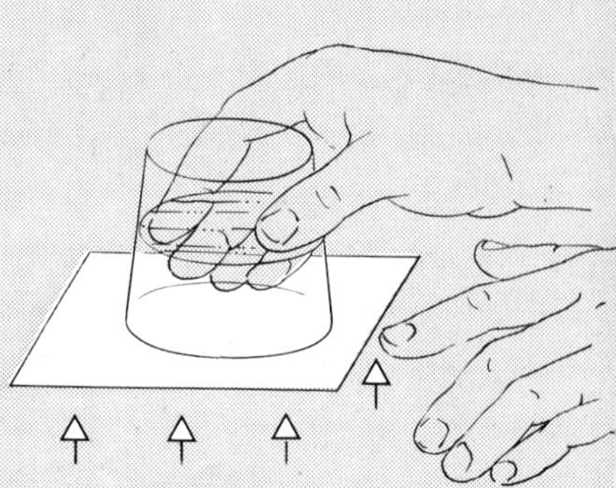

Take an ordinary drinking glass and half fill it with water. Then cover the glass with a thin piece of cardboard. This should be flat so that it can seal the glass.

Turn the glass upside down while holding the cardboard in place. This is best done over a sink, as there is always the possibility of something going wrong.

Now let go of the cardboard. Assuming there is no leak between the cardboard and glass, air pressure will support the cardboard and the water above it.

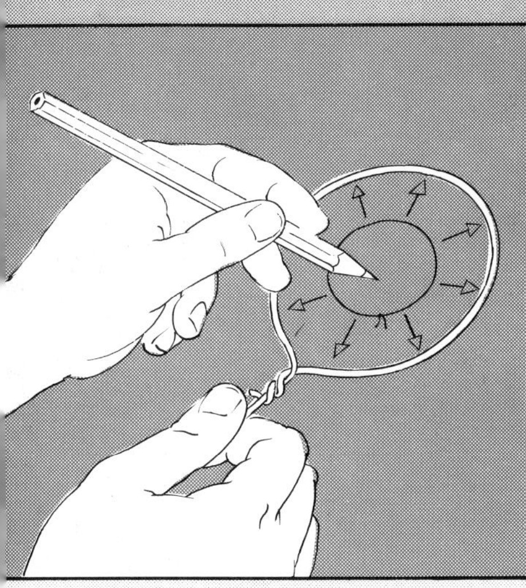

Now break the film inside the cotton. The surface tension in the remaining outer film will pull the cotton loop so that it forms a circle.

AIR PRESSURE

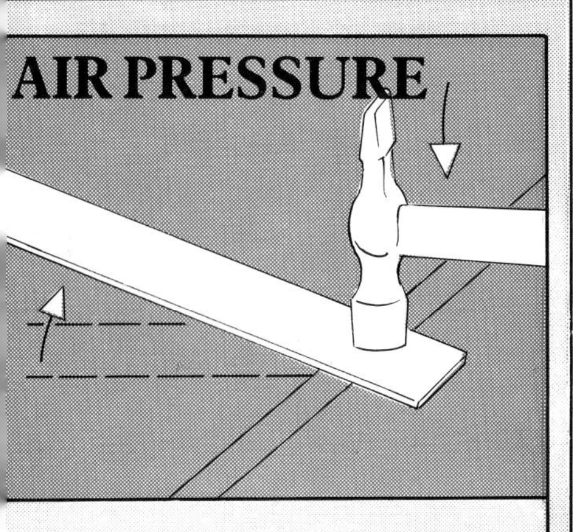

It is almost impossible to break a piece of wood like this. It merely jumps up in the air when you strike it. Pressure must be applied to resist this movement.

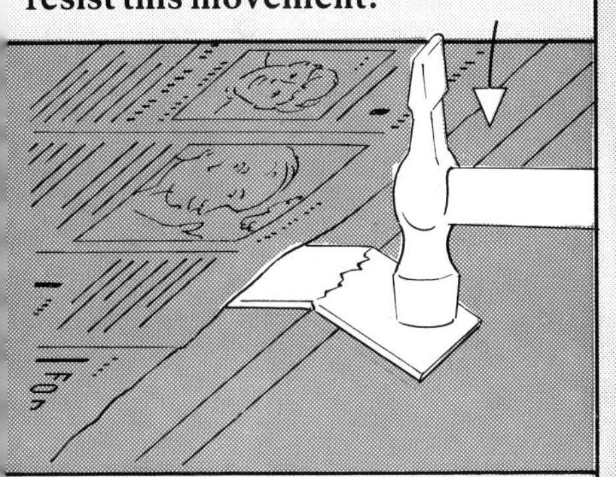

A sheet of newspaper, smoothed down, can provide the resistance. Air pressure acting on the paper tends to hold it down, so that the stick breaks when struck.

WHAT MAKES IT MOVE?

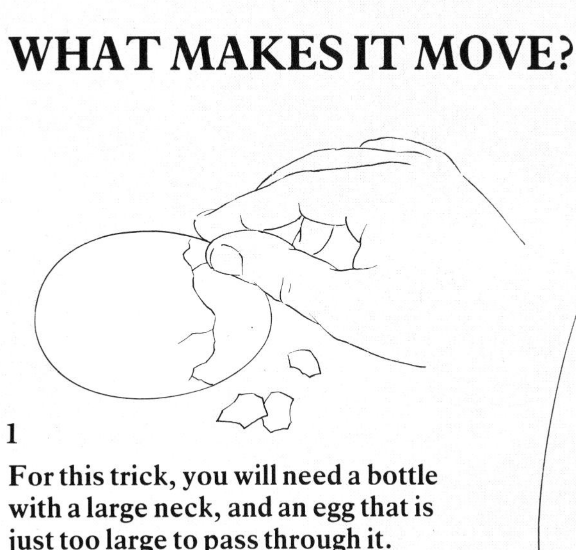

1

For this trick, you will need a bottle with a large neck, and an egg that is just too large to pass through it. Boil the egg for six minutes, let it cool, and then remove its shell.

2

Make sure that the bottle is quite dry inside. Now strike a few matches, dropping them into the bottle while they are still burning brightly.

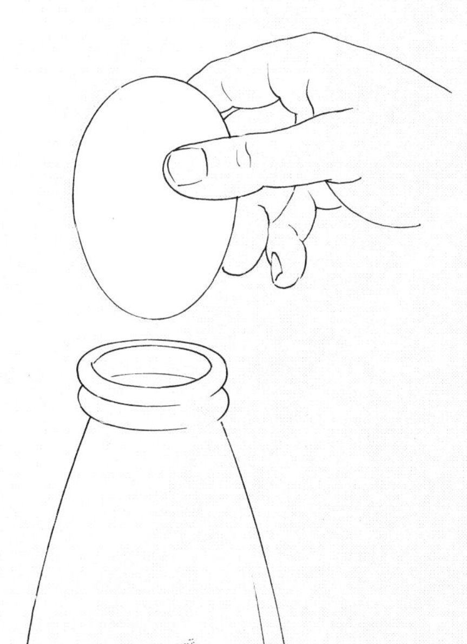

3

As the matches burn, they will heat the air in the bottle. When they have gone out, take the shelled egg and place it gently on top of the bottle.

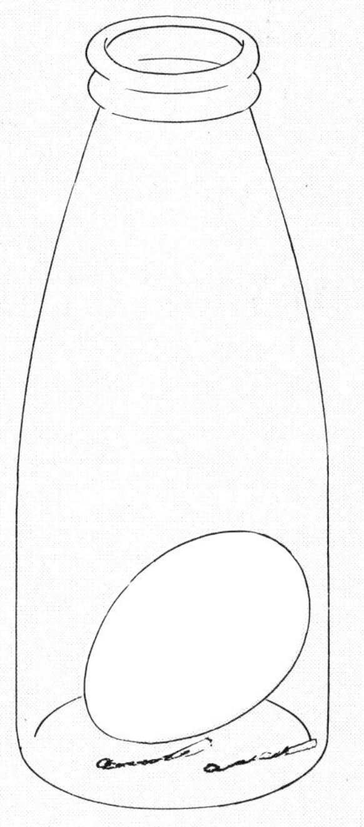

4

On cooling, the air inside reduces in pressure. The greater outside pressure then forces in the egg.

CHANCE AND LUCK

If you toss a coin and it comes down heads twice running, what are the chances of it coming down heads the next time? Some people would expect the run of heads to continue, while others would expect a tail next time 'because of the law of averages'. Who is right? The first project shows you how to solve this problem by means of a simple experiment.

The second project investigates the fairground game of rolling a coin. You win if the coin comes to rest entirely within any of the squares marked on the board. But what are your chances?

When you throw two dice, some scores are much easier to get than others. This is shown in the racing game below.

HEADS OR TAILS?

Toss a coin and note whether it shows heads or tails. Do this at least 100 times. Almost certainly, your results will show several runs of heads and tails, but no throw is affected by what has gone before. After two heads, a head or tail are equally likely.

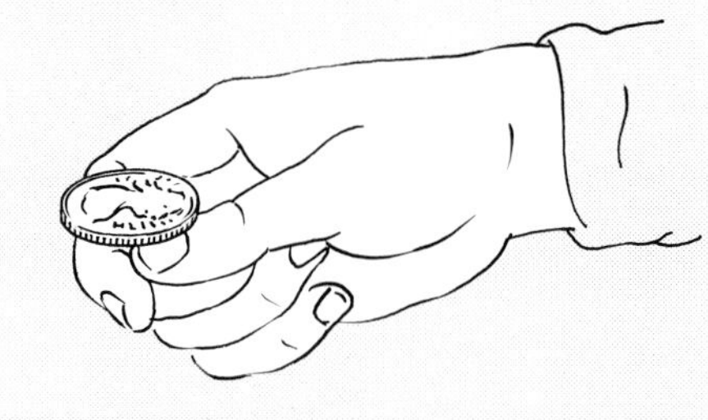

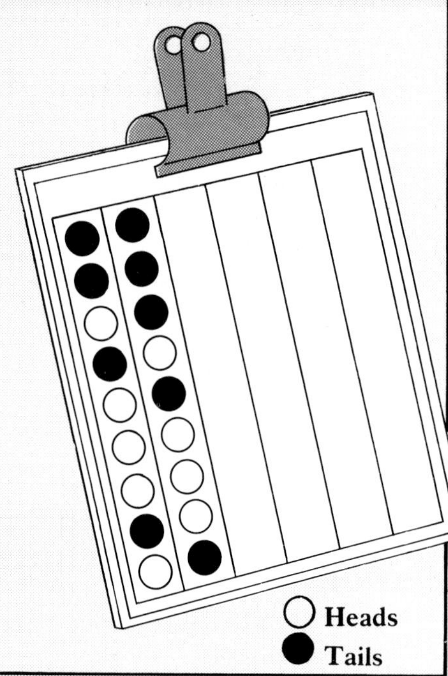

WHICH CAR STARTS FAVORITE?

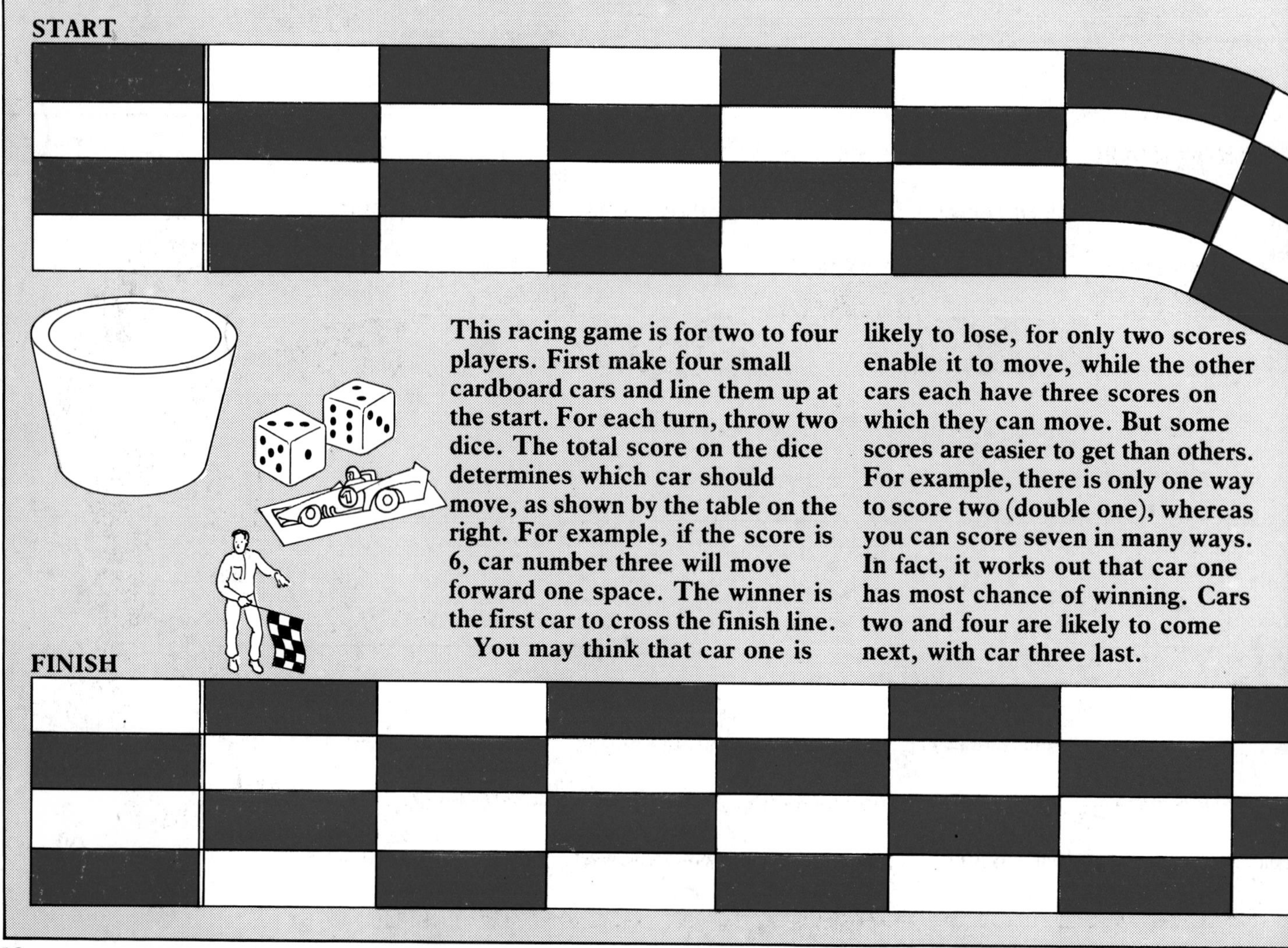

This racing game is for two to four players. First make four small cardboard cars and line them up at the start. For each turn, throw two dice. The total score on the dice determines which car should move, as shown by the table on the right. For example, if the score is 6, car number three will move forward one space. The winner is the first car to cross the finish line.

You may think that car one is likely to lose, for only two scores enable it to move, while the other cars each have three scores on which they can move. But some scores are easier to get than others. For example, there is only one way to score two (double one), whereas you can score seven in many ways. In fact, it works out that car one has most chance of winning. Cars two and four are likely to come next, with car three last.

ROLL-A-COIN

The roll-a-coin game is much more difficult than it appears. Even with fairly large squares, the coin often falls on a line. Using a piece of graph paper, you can quickly work out your chances of winning coin and squares of any size.

On a piece of graph paper, first draw a large square to represent a square on the board. Then draw a circle in a corner of the square to represent the coin. Be sure to use the same scale for the square and coin. Now draw a second square in the center of the first one, with one corner of the new square on the center of the circle. You win whenever the center of the coin is inside the smaller square. So your chance of winning is: area of small square divided by area of large square, here equal to 4 in 16, or 1 in 4.

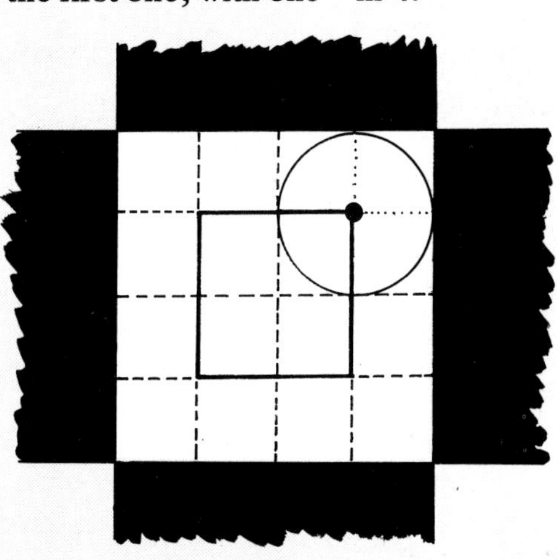

FINDING OUT MORE

Just because you have come to the end of this book, it need not be the end of your scientific experiments. In fact, it may be the start of a hobby that could develop into a career. Your local public library should be able to provide books and magazines on any particular subject that interests you. Here are a few ideas for expanding and applying some of the knowledge you have gained from this book.

Mind and Body Normal body temperature is said to be 98.4°F or 98.6°F, the latter being equal to exactly 37°C. Many people think that they must be ill if their temperature is slightly different from 'normal'. However, it is normal for our body temperature to vary. When is it at its highest, and when at its lowest? By how much does it vary? Does it vary by the same amount in different people? And why should it vary at all?

Keeping an Aquarium You can make a garden pond simply by digging a hole and lining it with thick polythene so that it will retain water. It is interesting to keep tadpoles and watch their gradual development into adult frogs.

Magnetism Magnets have many useful applications. For example, if you drop a box of pins or small nails, it is much quicker to pick them up with a magnet than by hand. A magnet can also be used to remove metal splinters safely from a wound.

Radio Receivers The volume of sound produced by a crystal set is usually quite low. This is because the set relies entirely on the electrical energy picked up by the aerial wire. There is no amplification device to strengthen the signals. Better results can be obtained by connecting the crystal set to the input socket of an amplifier and listening on a loudspeaker instead of headphones. Better still, build a transistor radio from one of the many kits advertised in radio and electronics magazines.

Amateur Detective Other clues, besides fingerprints, may help convict a criminal. Minute traces of soil or plant parts, for example, may lodge in the clothes or stick to the shoes. Using a good microscope, see if you can guess the identity of tiny samples of various substances chosen by a friend.

Codes and Secret Messages Many secret messages, written years ago, are waiting to be decoded. When picture postcards became extremely popular early this century, some people wrote on them in code to prevent the mail carrier from reading the message. Morse, mirror writing, shorthand and various private codes were used. See if you can find any old cards with coded messages to decipher.

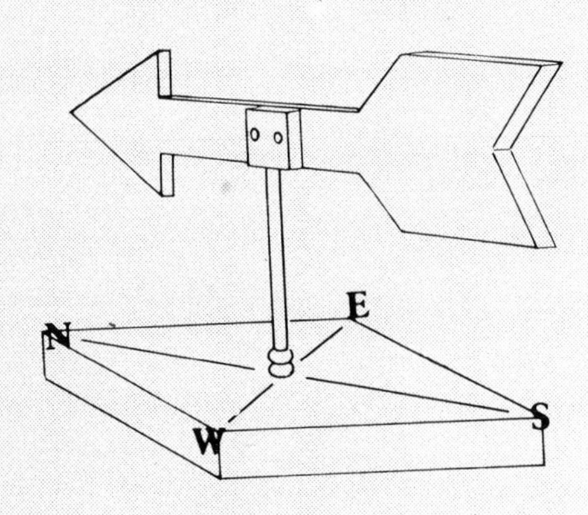

A Weather Station Some signs of coming changes in the weather can be obtained without special equipment. For example, if you listen to a radio tuned to the medium wave band, you will hear frequent interference if a thunderstorm is approaching. Some animals seem able to detect a coming storm and become restless.

Microscopes and Telescopes A wide range of instruments are available for those who require better results than can be obtained with homemade equipment.

Reflections Glue three mirrors to an inside corner of a box with their reflecting surfaces showing. You now have a special kind of reflector that will return a beam of light back to the source, regardless of the angle from which the beam comes. Where are such reflectors commonly used?

Optical Illusions You can use an ordinary camera to take 3-D photographs. Just take two pictures of a scene, moving the camera about an inch (a few centimeters) to one side between the shots. Place one of the resulting slides or prints (small) in front of each eye and view through two magnifying glasses. You will then have the illusion of seeing the view in three dimensions. You can exaggerate the three-dimensional effect by increasing the distance that you move the camera.

Making Things Move Model planes enable the experimenter to study the various factors, such as streamlining, that affect the performance of real aircraft. Models range from simple paper or balsa-wood gliders, through rubber-powered types to advanced, diesel-driven, radio-controlled aircraft.

Puzzles and Tricks There are many books on tricks, from simple conjuring to classic stage illusions. Anyone interested in puzzles can take their pick from the numerous regular magazines on this subject.

Chance and Luck Try to solve the following problem in your head. Then check your answer by experiment. Take three red and three black playing cards and place them face down in pairs thus: RR, RB, BB. Now ask a friend to select a card. If this is black, mix up all the cards and start again. If the chosen card is red, your friend wins if the second card is black. He loses if the second card is red. What are his chances of winning?

You will also want to read

MAKE YOUR OWN MUSICAL INSTRUMENTS

From a few tools, some pieces of wood, and odds and ends found around the house, you can make your own musical instruments. This book shows you how to create anything from a simple drum to a recorder or an eight-string zither. You can play your instrument by yourself or get together with friends to make a band. Discover the challenge and excitement of creating music with an instrument you made yourself!